STUDIES IN LINEAR PROGRAMMING

Studies in Management Science and Systems

Editor

Burton V. Dean
Department of Operations Research
Case Western Reserve University
Cleveland, Ohio

VOLUME 2

NORTH-HOLLAND PUBLISHING COMPANY – AMSTERDAM • OXFORD
AMERICAN ELSEVIER PUBLISHING COMPANY, INC. – NEW YORK

Studies in Linear Programming

Edited by

HARVEY M. SALKIN

Department of Operations Research
Case Western Reserve University
Cleveland, Ohio

and

JAHAR SAHA

Indian Institute of Management
Ahmedabad, India

1975

NORTH-HOLLAND PUBLISHING COMPANY – AMSTERDAM • OXFORD
AMERICAN ELSEVIER PUBLISHING COMPANY, INC. – NEW YORK

Library of Congress Catalog Card Number: 74 28998
North-Holland ISBN for the series: 0 7204 8700 5
North-Holland ISBN for this volume: 0 7204 8704 8
American Elsevier ISBN: 0 444 10884 x

Published by:

North-Holland Publishing Company – Amsterdam
North-Holland Publishing Company, Ltd. – Oxford

Sole distributors for the U.S.A. and Canada:

American Elsevier Publishing Company, Inc.
52 Vanderbilt Avenue
New York, N.Y. 10017

Printed in The Netherlands

To my wife
   Laura May Salkin

To my parents
   Jnanada and Hiralal Saha

CONTENTS

CONTENTS

PREFACE

This volume contains a collection of papers describing applications of linear programming in industrial, agricultural, and other environments that have appeared in this country and abroad. Topics range from classical refinery problems to more contemporary applications such as paper recycling. Each article gives the problem description, development of the mathematical model, and the experiences in using the model in practice. The exposition is given at a modest mathematical level. In addition, the first chapter is devoted to general applications of linear programming, a simple description of the simplex method, and a discussion of current linear programming codes and solution capabilities.

The book is intended for managers, operations researchers, engineers, systems analysts, students, and others who want to become familiar with the uses and potential uses of linear programming. It can serve as a reference, or as a textbook in a linear programming applications course for students of management and operations research (e.g., in a M.S. in Management, M.B.A., B.S. in Management, or in a similar program). The book is being used at Case in our Linear Programming and Applications course. Because of the extended use, interest, and importance of linear programming as a decision tool, and because of the broad range of applications which appear in the book, the volume can serve as a vehicle to acquaint any individual involved in management decisions with the efficacy of linear programming.

Each article has been carefully refereed and edited. We have tried to make the organization of each chapter somewhat uniform. In particular, chapters usually contain sections titled Introduction, The Problem, The Model, and Computational Results. Variables are usually denoted by the capital letters U, V, W, X, Y, and Z. Other symbols are usually constants.

As already mentioned, the first article presents an exposition to linear programming and related areas. Part I, comprising Chapters 2 through 7, contains articles describing situations resulting from an industrial environment. Topics include corporate planning, petroleum distribution, lease-buy planning,

production scheduling, and financial management situations. The next part describes problems resulting from agricultural situations. In particular, Chapter
8 describes a farm expansion planning problem. This is followed by an article
presenting a cattle feedlot-recycling scenario, while Chapter 10 contains a
macro-agricultural planning model. Part III embodies other situations which result in linear programming models. Chapters 11 through 15 include problems
from tourist investment planning, river basin planning, transportation planning,
and community development.

   Although each article presents an interesting application of linear programming some are naturally easier to read and understand than others. This is
mainly because the problem situations contain varying degrees of complexity.
Therefore, if the reader is not familiar with linear programming and/or is not
interested in any particular application, we suggest that the easier chapters be
read first. The diagram below gives our opinion as to general complexity and
readability of each article. Chapter 1 is not listed as it is a necessary back-
ground article for those not familiar with linear programming.

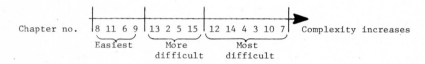

   As this volume suggests, many scenarios can be modeled as a linear program.
However, as the complexity of the situation increases, the number of assumptions
necessary to the model formulation and the difficulty in obtaining reliable data
tend to dramatically increase. Moreover, the solution of the resulting model may
be a difficult task, requiring a substantial amount of time, effort, and money.
If the linear program can be solved using the available resources the results
are, in numerous instances, used only for planning and guideline purposes. That
is, as additional information to support other related decisions. This typical
use of linear programming will undoubtedly continue. (Naturally, not all situations conform to the above. For example, Chapter 6 reflects a model drawn from a
relatively sterile environment with virtually error-free data. The results, as
dictated by the model, have been used.)

We are indebted to many individuals for making this volume possible. The contributing authors are sincerely thanked for writing about their work and for their patience and cooperation during the editing process. We are also grateful to Burton V. Dean, the series editor for the inclusion of this volume and general encouragement. Our appreciation is also extended to Sue Ramona, Sue Preston, and Barbara Bensch for typing the manuscript.

Cleveland, Ohio   H.M.S.
January 1975   J.S.

Chapter 1

AN INTRODUCTION TO
LINEAR PROGRAMMING

Harvey M. Salkin
Department of Operations Research
Case Western Reserve University
Cleveland, Ohio 44106

and

Jahar Saha
Indian Institute of Management
Ahmedabad, India

Abstract. An introduction to linear programming,
including its uses, importance, and computations,
is presented. Also, the mathematical theory used
in developing solution algorithms is briefly outlined and some techniques are given. Available
linear programming computer codes with capability
estimates are listed.

1. *Introduction*

A *management science* (or *operations research*) approach to a problem situation
may appear as in Figure 1. After a problem situation is defined, its composition
is analyzed, and the resulting components are represented by symbols and related
using equations, forming a *mathematical model*. The model is then solved using some
technique. If the results appear reasonable and acceptable to management, they
may be used in some fashion (e.g., for planning); otherwise, the problem and/or
model may have been described and/or formulated incorrectly, and a return to an
earlier stage is necessary. In passing from the component stage to the model
formulation stage, various, hopefully reasonable, assumptions are made. In fact,
these are often necessary to ensure that the problem can be posed via a mathematical description and the resulting model can be solved at an acceptable expense.

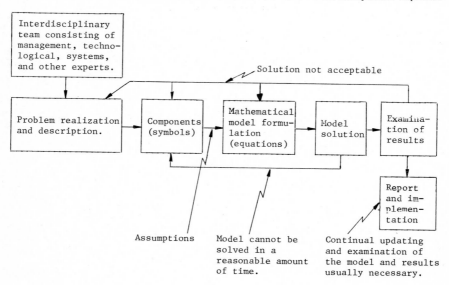

FIGURE 1. A MANAGEMENT SCIENCE APPROACH

A *system* can be thought of as a set of interrelated *activities*. Many real
world situations are of the sort where we seek to determine the level of these
activities so that the system performs in a best or *optimal* way. Usually, there
are various factors constraining the activity levels. The equation measuring the
system's performance is called the *objective function* and the activities are
labelled *variables*. An optimization model, where we wish to find a set of nonnegative values for the variables subject to the constraints, while obtaining a
best value for the objective function, is called a *mathematical programming
problem*. (See Figure 2, solid arrows.)

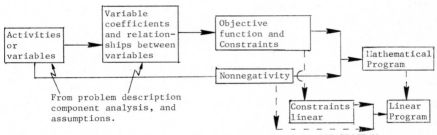

FIGURE 2. COMPOSITION OF A MATHEMATICAL PROGRAM AND A LINEAR PROGRAM

Although many scenarios yield mathematical programs, few of them can be
solved in a reasonable amount of computer time. This is because the mathematical
theory surrounding the related techniques is relatively weak. The only clear
exception to this is a special kind of mathematical program called a *linear program*. A linear program contains an objective function which is a linear term and
all constraints are linear as well (see Figure 2).[1] If a problem situation can be
formulated, using reasonable assumptions, as a linear program, then it can be
solved. Current capabilities allow for the solution of linear programming models
with virtually any number of variables and thousands of constraints. This empha-
sizes their importance. A few simple scenarios which are modelled as linear programs are now given. Solution techniques are discussed in subsequent sections.

Example 1. (A transportation problem)

There are n customers (or destinations) each with a known demand b_j (j=1,...,n)
for a single product which can be produced at any of m known locations (or

---

[1] Loosely speaking, a term is called linear if and only if it is the sum or difference of components, where each component is a known constant or a variable times
a known constant. For example, the inequality $3X_1+2X_2 \leq 4$, where X_1 and X_2 are
variables is a linear inequality; while the inequality $3X_1^2+2X_2 \leq 4$ or $3X_1X_2+2X_2 \leq 4$
(etc.) is not. Note that components involving variables are usually written on
the left and the constants appear on the right hand side.

origins). Each production center has a known manufacturing limit a_i (i=1,..,m).
The per unit shipping cost from each origin i to destination j, denoted by c_{ij},
is known. The problem is to find the number of units to ship from each production center i to customer j so that capacities are not exceeded and all demands
are met at a minimal total transportation cost.

To model this problem as a linear program, we must first define the variables
or the level of activities we wish to determine. In this case, they are the number of units of each product to send from each origin i to destination j. Thus,
we shall let X_{ij} be the number of units to send from origin i to destination j.
As i=1,..,m and j=1,..,n, there are m x n variables, and by definition, $X_{ij} \geq 0$
for all i and j.

Let us consider the supply constraints first. Consider any origin i with a_i
available units. These units can be sent to any of the n customers. The number
of units sent to customer 1 is X_{i1}, to customer 2, X_{i2},...., to customer n, X_{in}.
Thus, the total number of units sent from origin i to all the customers is
$X_{i1} + X_{i2} +...+X_{in}$, or

$$\sum_{j=1}^{n} X_{ij};$$

and to ensure that the supply is not exceeded, we have

$$\sum_{j=1}^{n} X_{ij} \leq a_i$$

for each origin i (i=1,...,m) (Figure 3a).

Similarly, at any customer or destination j, b_j units are required. These
can be received from any of the production locations i (i=1,...,m). The number
received from origin 1 is X_{1j}, from origin 2, X_{2j},...., from origin m, X_{mj}. Hence,
the total number of units received at customer j from all the origins is
$X_{1j} + X_{2j} +...+ X_{mj}$, or

$$\sum_{i=1}^{m} X_{ij}.$$

Thus we have the constraints

$$\sum_{i=1}^{m} X_{ij} \geq b_j$$

for each customer j=1,...,n (Figure 3b).

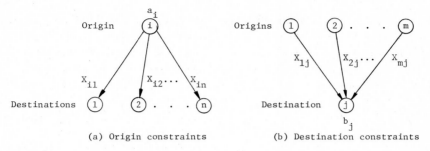

(a) Origin constraints   (b) Destination constraints

FIGURE 3.   SHIPPING CONSTRAINTS

The cost of shipping one unit between any origin i and destination j is c_{ij}. The number of units shipped is X_{ij}. Therefore, the shipping cost between any origin i and destination j is $c_{ij}X_{ij}$; and the total shipping cost from all origins to all destinations is

$$\sum_{i=1}^{m} \sum_{j=1}^{n} c_{ij}X_{ij}.$$

As we wish to minimize this quantity, the objective function is

minimize $$\sum_{i=1}^{m} \sum_{j=1}^{n} c_{ij}X_{ij}.$$

Summarizing, we have the linear program:

Minimize $\sum_{i=1}^{m} \sum_{j=1}^{n} c_{ij}X_{ij}$ Objective function

subject to $\sum_{j=1}^{n} X_{ij} \leq a_i$ $(i=1,\ldots,m)$ $\Bigg\}$

                $\sum_{i=1}^{m} X_{ij} \geq b_j$ $(j=1,\ldots,n)$ Constraints

and $X_{ij} \geq 0$ for all i,j.   Nonnegativity

Notice that a solution exists whenever the total supply is at least equal to the total demand. That is, whenever

$$\sum_{i=1}^{m} a_i \geq \sum_{j=1}^{n} b_j.$$

Moreover, if

$$\sum_{i=1}^{m} a_i = \sum_{j=1}^{n} b_j$$

and/or $c_{ij} > 0$, any solution will yield equalities in the constraints; and thus, in these cases, they can initially be taken as equations.

As a final note, consider any (e.g.) 2 origin and 3 destination illustration. The linear program is exhibited in Figure 4 in detached coefficient form. The matrix of constraint coefficients has special structure. In particular, all co-efficients are either 0 or 1 and each column has exactly 2 ones. Morever, if we denote the column of coefficients corresponding to the variable X_{ij} as P_{ij}, then a 1 appears in row i and in row m+j (m is the number of origins). For example, P_{21} has a 1 in row 2 and in row 3 (m=2). Because of this, it turns out that whenever the values of a_i and b_j are integral every optimal solution to the trans-portation problem, produced when solved as a linear program, will be integer. A careful examination of the initial scenario reveals that we must have the values of X_{ij} integer. However, if we had to enforce the constraint explicitly, the problem could no longer be posed as a linear program and solved using a linear programming algorithm.

X_{11}	X_{12}	X_{13}	X_{21}	X_{22}	X_{23}	Variables
c_{11}	c_{12}	c_{13}	c_{21}	c_{22}	c_{23}	Objective function
1	1	1	0	0	0	$\leq a_1$ Origin constraints
0	0	0	1	1	1	$\leq a_2$
1	0	0	1	0	0	$\geq b_1$ Destination constraints
0	1	0	0	1	0	$\geq b_2$
0	0	1	0	0	1	$\geq b_3$

P_{21}

Coefficient matrix

FIGURE 4. A 2 ORIGIN, 3 DESTINATION TRANSPORTATION PROBLEM

Example 2. (A product mix problem)

A company manufactures two products, A and B. Product A must be processed on three different machines while product B requires only two of them. The per unit time requirements for each product on each machine, and the total number of time units available for each machine, are tabulated below. Each unit of products A and B manufactured is sold for a net profit of $5 and $4, respectively. The problem is to determine the number of products of each type to manufacture so that the total net profit is maximized. Assume that there are no machine setup times involved.

		Time units required		Time units
		Prod. A	Prod. B	available
Machine	1	2	6	50
	2	8	0	40
	3	7	14	80

To model this situation as a linear program, we let X_1 and X_2 denote the number of units of product A and product B, respectively, to manufacture. Clearly X_1 and X_2 must be nonnegative. The time constraints below are also evident.

Machine	Constraint
1	$2X_1 + 6X_2 \leq 50$
2	$8X_1 \leq 40$
3	$7X_1 + 14X_2 \leq 70$

The total net profit is $5X_1 + 4X_2$. Or the linear program representing this situation may be written as below.

Maximize $\quad 5X_1 + 4X_2$

subject to $\quad X_1 + 3X_2 \leq 25$

$\qquad\qquad\quad X_1 \leq 5$

$\qquad\qquad\quad X_1 + 2X_2 \leq 10$

and $\qquad\quad X_1, X_2 \geq 0.$

Notice that the problem, as given, requires that the variables take on integer values. If fractional values for some of the variables appear subsequent to the linear program's solution, then they can be rounded to integer values. The resulting solution should be tested to see if all the constraints are satisfied or almost satisfied.[2] (If not, another rounding procedure can be tried.) Hopefully, a solution produced this way will be a near optimal one. The reader should realize, however, that generating solutions to linear programs with integer

---

[2] If the data reflects a reasonable amount of possible error, then slightly violated constraints may be considered satisfied and an "almost solution" may be quite acceptable.

constrained variables (called *integer programs*) in this manner may produce poor
solutions or none at all. A more complete discussion about rounding including
illustrative examples are in [6].

2. *Graphical Solutions*

Linear programs with at most 3 variables can be solved using a graph.
Although only the most trivial situations yield such modestly sized linear programs, the graphical technique is worth examining because it suggests several
results upon which the standard algorithm for solving linear programs, called
the *simplex method*, is based. We now present a few examples.

Example 3.   (A unique optimal solution)

Consider the linear program:

Maximize $5X_1 + 3X_2 = X_0$ (0)

subject to $3X_1 + 5X_2 \leq 15$ (1)

             $5X_1 + 2X_2 \leq 10$ (2)

and X_1 , $X_2 \geq 0$,   (3)

where X_0 is defined to be the value of the objective function $5X_1 + 3X_2$.
Suppose we define a point (X_1, X_2) which satisfies the constraints ((1) and (2))
as a *solution*, and a nonnegative solution as a *feasible solution*. A feasible
solution which gives X_0 its maximum value will be called an *optimal solution*.
Thus, in this example, an optimal solution (X_1, X_2) satisfies (1), (2), and (3)
while giving $5X_1 + 3X_2$ its largest value. To find an optimal solution, let us
first consider the set of feasible solutions, called the *feasible region*. By
plotting in (X_1, X_2) space (Figure 5a) the set of all points satisfying (1), (2),
and (3) we can visualize the feasible region. To do this, notice that the nonnegativity (3) simply restricts our graph to the first quadrant. To find the set
of (X_1, X_2) satisfying (1), consider the equation $3X_1 + 5X_2 = 15$. It describes a
straight line, and as two points determine a line, we can easily plot it on the
graph. For example, when $X_1 = 0$, $X_2 = 3$ (15/5), and when $X_2 = 0$, $X_1 = 5$ (15/3).
Connecting these two points (namely, $(X_1, X_2) = (0, 3)$ and $(5, 0)$) yields the
equation $3X_1 + 5X_2 = 15$. As this line separates the graph into two regions,
where in one $3X_1 + 5X_2 \leq 15$ and in the other $3X_1 + 5X_2 \geq 15$, the set of points
satisfying (1) can be found by picking a point in one of the regions. If it
satisfies (1) then that region contains all points which satisfy $3X_1 + 5X_2 \leq 15$;
otherwise (1) defines the set of points in the other region. For example, when
$X_1 = X_2 = 0$, we have $3(0) + 5(0) \leq 15$ so that (1) defines all points on and

below the line (see Figure 5b).

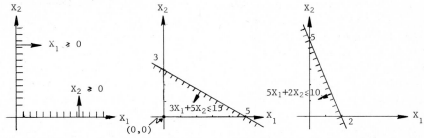

(a) The first quadrant   (b) The first constraint   (c) The second constraint

FIGURE 5.   FINDING THE FEASIBLE REGION

The set of (X_1, X_2) satisfying (2) can be found in an analogous manner. That is, first the equation $5X_1 + 2X_2 = 10$ is plotted on the graph using (e.g.) the two points $X_1 = 0$, $X_2 = 5$ and $X_2 = 0$, $X_1 = 2$. Then a point (e.g.) $X_1 = X_2 = 0$ is tested, and as $5(0) + 2(0) < 10$ the constraint yields the points as in Figure 5c.

To find the feasible region we merely plot (1), (2), and (3) on the same graph as in Figure 6. This diagram can be thought of as an aggragate of those in Figure 5 and yields the intersection of the regions indicated by the border.

An optimal solution is one of the feasible solutions and is thus a member of the feasible region. Hence, we might try to find it by selecting various values for X_1 and X_2 from the feasible region and computing $X_0 = 5X_1 + 3X_2$. Naturally,

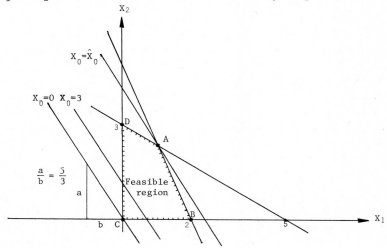

FIGURE 6.   THE FEASIBLE REGION

the trouble with this approach is that there are an infinite number of feasible solutions in this example. However, by fixing X_0, notice that $5X_1 + 3X_2 = X_0$ is a straight line which can be plotted on the graph. For example, when $X_0 = 0$, we have $5X_1 + 3X_2 = 0$ which appears in Figure 6. Taking a second value, say $X_0 = 3$, yields $5X_1 + 3X_2 = 3$ and another line in the diagram. Notice that the lines $X_0 = 0$ and $X_0 = 3$ appear parallel, and in fact they are. A careful examination of the objective function reveals that the slope is always $-5/3$ and independent of the value of X_0.[3] Therefore, since X_0 increases when the objective function moves deeper in the first quadrant, the optimal solution can be found by pushing the line $5X_1 + 3X_2 = 0$ up and parallel to itself until it will no longer touch the feasible region. The last member of the feasible region on the objective function line is optimal. In this case it is point A in Figure 6.

To find the components (X_1, X_2) of point A, notice that they satisfy the equations $3X_1 + 5X_2 = 15$ and $5X_1 + 2X_2 = 10$. Solving this system of two equations in two unknowns yields $X_1 = 20/19$ and $X_2 = 45/19$, or A = $(20/19, 45/19)$. The maximum value of X_0, denoted by $\hat{X}_0$, is found using $\hat{X}_0 = 5X_1 + 3X_2 = 5(20/19) + 3(45/19) = 235/19$.

The previous discussion suggests that an optimal solution to a linear programming problem occurs at a "corner point" of the feasible region. (For example, in Figure 6 the four heavily dotted points, labelled A, B, C, D, are corner points.) In fact, this is always true. A corner point, appearing at a vertex of the feasible region, is called an *extreme point*. To define an extreme point more precisely, we first define a *convex set* as a set in which the line segment joining any two points in the set lies wholly in the set. Figures 7a, 7b, and 7c display convex sets while the other diagrams in the figure depict sets which are not convex. Any point which can be removed from a convex set, while the set remains convex is an extreme point. Notice that the curved surfaces contain extreme points along their perimeter. It turns out (see [1], [4]) that the feasible region is always a convex set and an optimal solution occurs at an extreme point. However, as the next example illustrates, there may be optimal solutions which are not extreme points.

---

[3] A well known result from geometry ([8]) is that given the line $y = mx+b$, where x describes the horizontal axis and y the vertical axis, the value of m is the slope and b is the y intercept. The slope is defined to be the tangent of the angle between the line and the horizontal (or x) axis. In our case, $x = X_1$, $y = X_2$, and $y = mx+b$ is equivalent to $X_2 = (-5/3)X_1 + X_0/3$ or $m = -5/3$ regardless of the value of X_0.

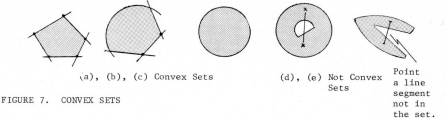

(a), (b), (c) Convex Sets (d), (e) Not Convex Sets Point a line segment not in the set.

FIGURE 7. CONVEX SETS

Example 4. (Alternate optimal solutions)

Suppose we have the same feasible region as in the previous example but the objective function is now: Maximize $(5/2)X_1 + X_2 = X_0$. Then we have Figure 8 copied from Figure 6, except that the new objective function is included. As its slope is $-5/2$, the objective function is parallel to the constraint equality (2), namely, $5X_1 + 2X_2 \leq 10$. The extreme points $A = (20/19, 45/19)$ and $B = (2, 0)$ are optimal, as well as all points on the line segment joining A and B (see Figure 8). The maximum value of X_0, $\hat{X}_0 = 5/2(2) + 0 = 5/2(20/19) + 45/19 = 5$.

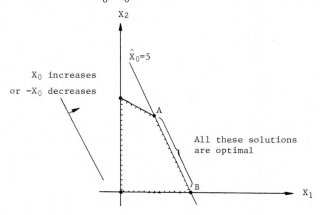

FIGURE 8. ALTERNATE OPTIMAL SOLUTIONS

When, as in the previous example, there is more than one optimal solution, we say there are *alternate optima* or *alternate optimal solutions*. Also, note that had we solved the previous example with the objective function: Minimize $(-5/2)X_1 - X_2 = \bar{X}_0 = (-X_0)$, in place of Maximize $(5/2)X_1 + X_2 = X_0$, then exactly the same set of optimal solutions would be found with the minimum of $\bar{X}_0 = -5$. This means that we can always solve a linear program as a maximization problem by simply multiplying the objective function in a minimization problem by -1.

Other examples, illustrating different possibilities are given below.

Example 5. (Solution unbounded)

Suppose the linear program produces one of the situations portrayed in Figure 9. Then we say the *solution is unbounded*. That is, there is no optimal solution. If this occurs the model may have been formulated incorrectly and certain resource constraints may have been inadvertantly omitted.

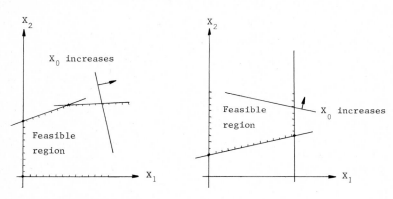

FIGURE 9.   SOLUTION UNBOUNDED ($X_0 \to +\infty$)

Example 6. (No feasible region)

The feasible region may not exist if the constraints are inconsistant in which case there are no solutions (Figure 10a), or there are no feasible solutions although solutions may exist (Figure 10b). In either case, the model or problem may have been formulated incorrectly.

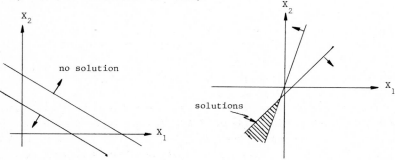

(a)   Inconsistant constraints   (b)   No feasible solution

FIGURE 10.   NO FEASIBLE REGION

From the examples we can see that a linear program either has an optimal
solution, an unbounded solution, or no feasible solution. Moreover, if it has
an optimal solution, at least one will occur at an extreme point of the feasible
region which is always a convex set. In the next section we will explain the
equivalance between certain feasible solutions and extreme points. This will
allow us to develop an iterative algorithm which explores adjacent or neighboring
extreme points so that no one extreme point is examined indefinitely and the
value of X_0 will never decrease. As it can be shown that the number of extreme
points is finite (see [1], [4]), this will mean that the technique converges to
an optimal extreme point. The method will also show when the linear program has
no feasible solution or an unbounded solution. The algorithm, called *the simplex
method,* is now discussed.

3. *The Simplex Method*

The algorithm will be presented using Example 3, which is written below.

Maximize $\quad 5X_1 + 3X_2 = X_0 \qquad$ (0)

subject to $\quad 3X_1 + 5X_2 \leq 15 \qquad$ (1)

$\qquad\qquad\quad 5X_1 + 2X_2 \leq 10 \qquad$ (2)

and $\qquad\quad X_1, X_2 \geq 0. \qquad$ (3)

Putting the System in Canonical Form

The first step is to pose each inequality constraint, except the nonnegativity
one (3), as equations with a nonnegative right hand side term. This is accomplished by first multiplying each constraint with a negative right hand side value
by −1 so that it becomes positive. Naturally, inequality signs are reversed upon
this multiplication. In our example, both right hand side terms, 15 and 10, are
nonnegative so nothing has to be done. To convert inequality constraints to
equations we adjoin a new nonnegative variable to each inequality. In particular,
in $3X_1 + 5X_2 \leq 15$, define (say) $X_3 = 15-(3X_1 + 5X_2)$, where $X_3 \geq 0$ comes from
inequality (1). Then rearranging terms, putting all variables on the left hand
side, yields $3X_1 + 5X_2 + X_3 = 15$. Similarly, defining $X_4 = 10-(5X_1 + 2X_2)$ in
(3), we have $5X_1 + 2X_2 + X_4 = 10$ and $X_4 \geq 0$. Thus, expressions (1), (2), and
(3) become

$$3X_1 + 5X_2 + X_3 \quad\;\; = 15 \qquad (1)'$$
$$5X_1 + 2X_2 \quad\;\; + X_4 = 10 \qquad (2)'$$
$$X_1, X_2, X_3, X_4 \geq 0. \qquad\quad (3)'$$

Notice that had we a constraint inequality of the sort $-2X - 3X_2 \leq -1$, we would first multiply it by -1 to obtain $2X_1 + 3X_2 \geq 1$ and then define the new variable (say) $X_5 = 2X_1 + 3X_2 - 1$ so that $X_5 \geq 0$. Rearranging terms produces $2X_1 + 3X_2 - X_5 = 1$. Summarizing we have that, once each constraint has a nonnegative right hand side term, inequalities of the $\leq$ sort become an equation by adding a new nonnegative variable with a $+1$ coefficient, and those inequalities of the $\geq$ type are converted to equations by adjoining a new nonnegative variable with a -1 coefficient. In any case, the new variable is adjoined to the left hand side. The variables with a $+1$ coefficient are termed *slack variables* and those with -1 coefficients as *surplus variables*.

In the next initialization phase, the objective function (4) is written as $X_0 - 5X_1 - 3X_2 = 0$ and listed with the constraints. The entire problem is thus equivalent to:

Maximize X_0

subject to $X_0 - 5X_1 - 3X_2 \qquad\qquad = 0 \qquad (0)'$
$$3X_1 + 5X_2 + X_3 \quad\;\; = 15 \qquad (1)'$$
$$5X_1 + 2X_2 \quad\;\; + X_4 = 10 \qquad (2)'$$

and $X_1, X_2, X_3, X_4 \geq 0. \qquad (3)'$

At this point we relate the above system of equations $(1)'-(3)'$ to an extreme point of the feasible region[4] by examining each equality. In particular, we inspect each equation, except the one representing the objective function (i.e., $(0)'$) for a variable which appears only in that equation and has a $+1$ coefficient. If such a variable appears for *each* equation we say that the system is in *canonical form*. In our case, we have that X_3 satisfies the requirements for $(1)'$ and X_4 for $(2)'$, so that the system of equations $(0)'-(2)'$ is in canonical form. Variables such as these are called *basic variables*. Notice that X_0 always

---

[4] It can be shown ([1], [3]) that there is a one to one correspondence between the feasible region, and hence extreme points, defined by the original constraints and the one produced by the constraints converted to equation form. In the example, (X_1, X_2) is an extreme point of the region defined by (1)-(3) if and only if (X_1, X_2, X_3, X_4) is an extreme point of the region produced by $(1)'-(3)'$.

satisfies the requirements of a basic variable for the equation representing the
objective function, and thus for the first equation the basic variable is always
X_0. All other variables, in our case X_1 and X_2, are *labelled nonbasic variables*.
Observe that for a system with m equations there are exactly m basic variables.
In the example, m=3. If the system of equations is not in canonical form, we
explain what to do in the next section. (This will also take care of the case
when there is no feasible solution.)

Finding an Initial Basic Feasible Solution

It can be shown ([1], [3]) that a system of equations in canonical form with
a nonnegative right hand side (except for the value in the first equation) yields
an extreme point of the feasible region. In particular, the feasible solution
obtained by setting the basic variables to the right hand side and the nonbasic
variables to 0, called a *basic feasible solution,* is an extreme point of the
associated feasible region. In the example, $X_0 = 0$, $X_3 = 15$, $X_4 = 10$ (basic variables) with $X_1 = X_2 = 0$ (nonbasic variables) is a basic feasible solution and
thus an extreme point. To see this, the constraints (1)-(3) are plotted in Figure 6. In that diagram, $X_1 = X_2 = 0$ where $X_0 = 0$ is an extreme point of the
feasible region.

Test for Optimality

Once an initial basic feasible solution has been found we must determine
whether it corresponds to an optimal extreme point. To do this we examine the
first equation of the system which is now in canonical form. In the example it
is

$$X_0 - 5X_1 - 3X_2 = 0. \qquad (0)'$$

Notice that, except for the +1 coefficient for X_0, only the nonbasic variables
can have nonzero coefficients in this equation. At the extreme point, the nonbasic variables, here X_1 and X_2, have value 0. If any one of them can be made
positive the value of X_0 will improve (i.e., increase) since the coefficient of
X_1 and X_2 in (0)' is negative. In particular, from (0)' we have that X_0 increases
by 5 each time X_1 is incremented by 1 and it increases by 3 everytime X_2 is
increased by 1. Therefore, the basic feasible solution exhibited in the equation
(0)'-(2)' may not correspond to an optimal extreme point. Moreover, had each
coefficient in the equation representing the objective function been nonnegative
then the current basic feasible solution is optimal. For example, had (0)' been

$$X_0 + 5X_1 + 3X_2 = 0 \qquad (0)''$$

then increasing X_1 or X_2 from value 0 would only make X_0 smaller or worse. In this case, the basic feasible solution $X_0 = 0$, $X_3 = 15$, $X_4 = 10$, and $X_1 = X_2 = 0$ is optimal. The objective function (0)" is: Maximize $-5X_1 - 3X_2$ which is equivalent to: Minimize $5X_1 + 3X_2$, and clearly if $X_1 = X_2 = 0$ is a member of the feasible region it minimizes $5X_1 + 3X_2$ as X_1 and X_2 must always be at least 0.

Finding an Improved Basic Feasible Solution

If the current basic feasible solution is not optimal, then it is possible to produce another one so that the value for X_0 does not worsen, and when it improves, a new neighboring extreme point is found. This is accomplished by exchanging a single nonbasic variable with a basic variable, and will be equivalent to producing a new system of equations in canonical form from the current system.

We have just established that by selecting a nonbasic variable with a negative coefficient in (0)', the value for X_0 cannot get worse. Furthermore, as the value of X_0 may increase only if a nonbasic variable with a negative coefficient in (0)' is increased, it seems natural to select the variable having the largest negative coefficient in absolute value to become a basic variable. This gives the greatest per unit increase in X_0.

The current basic variable which is to be exchanged and become nonbasic is found so that a basic feasible solution is maintained. We explain the calculations using the example.

Both X_1 and X_2 in (0)' have negative coefficients. As the coefficient of X_1, -5, is larger in absolute value than the coefficient of X_2, -3, it is selected to become a basic variable. The variable to become basic, i.e., X_1, is termed the *IN variable*.

To determine the current basic variable which is to be replaced by the IN variable, called the *OUT variable*, we find the maximum value that the IN variable can have so that a basic feasible solution is produced. To do this, the system in canonical form, except the equation representing the objective function, is examined. We have

$$3X_1 + 5X_2 + X_3 \qquad = 15 \qquad (1)'$$

$$5X_1 + 2X_2 \qquad + X_4 = 10. \qquad (2)'$$

Solving for the basic variables X_3 and X_4 in terms of the nonbasic ones gives

$$X_3 = 15 - 3X_1 - 5X_2 \qquad (1)''$$

$$X_4 = 10 - 5X_1 - 2X_2. \qquad (2)''$$

In the new system of equations in canonical form, X_2 will still be a nonbasic variable and thus set to value 0 in the new basic feasible solution. From (1)" and (2)", this means we must have

$$X_3 = 15 - 3X_1 \geq 0 \qquad (1)$$
$$X_4 = 10 - 5X_1 \geq 0 \qquad (2)'''$$

Equation (1)''' indicates that to have $X_3 \geq 0$, X_1 can be at most 5 (15/3) and to have $X_4 \geq 0$ equality (2)''' gives $X_1 \leq 10/5 = 2$. As both X_3 *and* X_4 must remain nonnegative, we have that X_1 can be at most the minimum of 5 and 2 or 2. Also, setting $X_1 = 2$ gives $X_3 = 9$ and $X_4 = 0$. As nonbasic variables are set to 0, this suggests making X_4 a nonbasic variable at the expense of making X_1 a basic variable; i.e., X_4 is the OUT variable.

Observe that the only candidates for the OUT variable are those whose corresponding equation has a positive coefficient for the IN variable. That is, in both (1)' and (2)' the coefficient of X_1 (the IN variable) is positive so that both X_3 and X_4 are candidates for the OUT variable. But suppose the coefficient of X_1 in (1)' had been -3 instead of $+3$, then (1)''' would have been $X_3 = 15 + 3X_1$. This means that X_1 can be made arbitrarily large while X_3 remains nonnegative. In this case, X_4 must be the OUT variable with (2)''' still giving $X_1 \leq 2$ (when $X_1 = 2$, $X_4 = 0$ and $X_3 = 15 + 3(2) = 21$). Had the coefficient of X_1 in (2)' also been nonpositive, say -5, then (2)''' is $X_4 = 10 + 5X_1$ and again X_1 can be made arbitrarily large while X_4 remains nonnegative. In this situation, the problem has an unbounded solution. In fact, it is of the form $X_1 = a$, $X_2 = 0$, $X_3 = 15 + 3a$, and $X_4 = 10 + 5a$, where as $a \to +\infty$, $X_0 = 5a \to +\infty$. Graphically solving the linear program

Maximize $5X_1 + 3X_2 = X_0$

subject to $-3X_1 + 5X_2 \leq 15$

$$-5X_1 + 2X_2 \leq 10$$

and $X_1, X_2 \geq 0,$

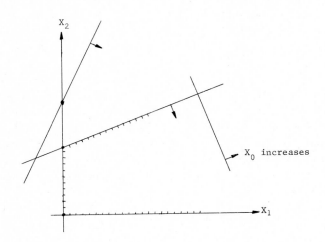

FIGURE 11. AN UNBOUNDED SOLUTION

To summarize, a new basic feasible solution can be found by determining which nonbasic variable is to become a basic one, termed the IN variable, and which basic variable is to become a nonbasic variable, termed the OUT variable. The IN variable corresponds to the nonbasic variable with the largest negative coefficient in absolute value in the equation representing the objective function. In every other equation, the coefficient of the IN variable is examined. If it is positive, the ratio of the associated right hand side value divided by this positive coefficient is computed. The basic variable corresponding to the equation that gives the minimum ratio is the OUT variable.[5] If every coefficient of the IN variable is nonpositive, then the problem has an unbounded solution.

Finding the New System in Canonical Form

To determine the new system of equations in canonical form and thus the next basic feasible solution, we have that X_0 and X_3 remain basic variables while X_1 is replacing X_4. As X_4 is in equation (2)', the coefficient of X_1 in this equation should be made +1 and X_1 should be eliminated from equations (0)' and (1)'.

---

[5] In case of ties, select any one. This could possibly, but not probably, mean that the algorithm will fail to converge or "cycle" (see footnote 6).

This can be accomplished by first dividing equation (2)' by 5, the coefficient of X_1, yielding equality $(\bar{1})$. Then, since the coefficients of X_1 in equations (0)' and (1)' are −5 and +3 respectively, adding +5 times equality $(\bar{2})$ to (0)' and −3 times $(\bar{2})$ to (1)', yields the desired system $(\bar{0})$-$(\bar{2})$.

$$X_0 \quad - \quad X_2 \quad + \quad X_4 = 10 \qquad (\bar{0}) = (0)' + 5(\bar{2})$$

$$+ \ (19/5)X_2 + X_3 - (3/5)X_4 = \ 9 \qquad (\bar{1}) = (1)' - 3(\bar{2})$$

$$X_1 + \ (2/5)X_2 \quad + (1/5)X_4 = \ 2. \qquad (\bar{2}) = (1/5)(2)'$$

As the coefficient of X_2 in equality $(\bar{0})$ is negative, the basic feasible solution $X_0 = 10$, $X_3 = 9$, $X_1 = 2$, and $X_2 = X_4 = 0$ may not correspond to an optimal extreme point and the process must be repeated. In particular, since −1 is the sole negative coefficient in $(\bar{0})$, X_2 is the IN variable. The minimum of $(1/1)/(19/5)$ and $(2/1)/(2/5)$ is $(9/1)/(19/5)$, or X_3 is the OUT variable. The new system of equations with basic variables X_0, X_2, and X_1 is below.

$$X_0 \quad + \quad X_3 + \ (2/5)X_4 = 235/19 \qquad (\bar{\bar{0}}) = (\bar{0}) + (\bar{\bar{1}})$$

$$X_2 + (5/19)X_3 - (3/19)X_4 = \ 45/19 \qquad (\bar{\bar{1}}) = (5/19)(\bar{1})$$

$$X_1 \ - \ (2/5)X_3 + (5/19)X_4 = \ 20/19. \qquad (\bar{\bar{2}}) = -(2/5)(\bar{\bar{1}})$$

Since all coefficients in equality $(\bar{\bar{0}})$ are nonnegative, we have that the basic feasible solution $X_0 = 235/19$, $X_1 = 20/19$, $X_2 = 45/19$, and $X_3 = X_4 = 0$ gives an optimal extreme point. This agrees with Example 3 and Figure 6.

To correlate the computations with the geometry, consider the feasible region portrayed in Figure 6. The association is tabulated below and indicates the successive examination of neighboring extreme points.

System of Equations in Canonical Form	Basic Feasible Solution					Extreme Point (see Figure 6)
	X_1	X_2	X_3	X_4	X_0	
(0)', (1)', (2)'	0	0	15	10	0	C
$(\bar{0})$, $(\bar{1})$, $(\bar{2})$	2	0	9	0	10	B
$(\bar{\bar{0}})$, $(\bar{\bar{1}})$, $(\bar{\bar{2}})$	20/19	45/19	0	0	235/19	A

Note that in this example each basic feasible solution corresponded to a different extreme point with X_0 successively improving. In most cases this will happen and thus, since the number of extreme points is finite, the technique must converge to an optimal extreme point. However, in certain situations it is possible to produce more than one basic feasible solution corresponding to the

same extreme point with X_0 naturally remaining constant.[6] In practice this
presents no difficulty and, using the technique as given here, eventually an
improved basic feasible solution corresponding to a neighboring extreme point
will be found;[7] and thus the algorithm will converge anyway.

The Simplex Tableau

To simplify the computations, the problem can be placed in a tableau. The
system of equations in canonical form appear row-wise in detached coefficient
form. The tableaux below represent a compact version of the previous calculations. The circled element, called the *pivot*, is the coefficient of the IN
variable in the row corresponding to the OUT variable.

Basic variables	X_0	IN X_1	X_2	X_3	X_4	Basic feasible solution
X_0	1	-5	-3	0	0	0
X_3	0	3	5	1	0	15
OUT X_4	0	(5)	2	0	1	10

$$\left. \begin{array}{l} X_0 = 0 \\ X_1 = X_2 = 0 \end{array} \right\} C$$

Basic variables	X_0	X_1	IN X_2	X_3	X_4	Basic feasible solution
X_0	1	0	-1	0	1	10
OUT X_3	0	0	(19/5)	1	-3/5	9
X_1	0	1	2/5	0	1/5	2

$$\left. \begin{array}{l} X_0 = 10 \\ \mathbf{X_1} = 2, \ X_2 = 0 \end{array} \right\} B$$

---

[6] This can only occur when one or more of the basic variables has value 0, or
when there is a degenerate basic feasible solution.

[7] To rigorously prove that *cycling*, or remaining at the same not optimal extreme
point indefinitely, cannot occur, additional rules, called anti-cycling or
degeneracy procedures, must be used when selecting the OUT variable (see
[1], [3]).

Basic variables	X_0	X_1	X_2	X_3	X_4	Basic feasible solution
X_0	1	0	0	1	2/5	235/19
X_2	0	0	1	5/19	-3/19	45/19
X_1	0	1	0	-2/5	5/19	20/19

Optimal

$X_0 = 235/19$
$X_1 = 20/19$ } A
$X_2 = 45/19$

Alternate Optimal Extreme Point Solutions

Alternate optimal extreme point solutions are exhibited by an optimal simplex tableau with a 0 coefficient in the X_0 row corresponding to a *nonbasic* variable. Designating that variable the IN variable and computing the OUT variable in the usual way produces another optimal extreme point. This is evident since the X_0 row, hence the value of X_0, does not change. To illustrate, had the final simplex tableau in the last example had a 0 coefficient for X_3 in the X_0 row (in place of the 1 coefficient), the X_2 variable can be replaced by X_3. This would produce the alternate optimal extreme point B in Figure 8.

Listing of the Simplex Algorithm

Suppose a tableau has the form below. Then the simplex algorithm can be listed.

Basic variables	X_0	X_1	. . .	X_n	Basic feasible solution
X_0	1	a_{01}	. . .	a_{0n}	b_0
X_{B1}	.	a_{11}	. . .	a_{1n}	b_1
.	.	.	.	.	.
.	.	.	.	.	.
.	.	.	.	.	.
X_{Bm}	0	a_{m1}	. . .	a_{mn}	b_m

Step 1. Place the original constraints in canonical form with nonnegative values for the basic variables, (i.e, $b_1 \geq 0,\ldots, b_m \geq 0$). Use artificial variables if necessary. (These are explained in the next section.) Go to Step 2.

Step 2. Let the IN variable X_k be found using $a_{0k} = \underset{j}{\text{maximum}} |a_{0j}|$ ($j \geq 1$), such that $a_{0j} < 0$. If $a_{0j} \geq 0$ for all j, the current basic feasible solution is optimal - terminate. Go to step 3.

Step 3. Find the OUT variable X_{Br} by b_r/a_{rk} = minimum b_i/a_{ik}, where $a_{ik} \geqslant 0$. If
all $a_{ik} \leqslant 0$ (i=1,2,...,m) the problem is unbounded – terminate. Go to step 4.

Step 4. Using a_{rk} as the pivot element, do a pivot operation and return to
step 2.

4. *Artificial Variables, Two Phase Methods*

The simplex method requires that an initial basic feasible solution be found,
or equivalently, that the constraints initially be converted to a system of equations in canonical form so that the right hand side is nonnegative. In our
example, the right hand side was nonnegative and a slack variable was added to
each equation so that the system was immediately in canonical form. However, in
many problems, posing the constraints in canonical form with a nonnegative right
hand side is not a simple task. To illustrate, consider the example below.

Example 7.

Maximize $-3X_1 - 4X_2 = X_0$ (3)

subject to $2X_1 - 2X_2 \leqslant -1$ (4)

$-2X_1 - X_2 \leqslant -2$ (5)

and $X_1, X_2 \geqslant 0.$ (6)

Multiplying each inequality by -1 and adding the surplus variable X_3 to (4) and
X_4 to (5) yields

$-2X_1 + 2X_2 - X_3 \quad\quad = 1$ (4)'

$2X_1 + X_2 \quad\quad - X_4 = 2$ (5)'

and $X_1, X_2, X_3, X_4 \geqslant 0.$ (6)'

As the coefficient of X_3 and X_4 is -1, *not* $+1$, these variables cannot serve as
basic variables. (This is evident since setting $X_1 = X_2 = 0$ yields $X_3 = -1$ and
$X_4 = -2$ which is not a feasible solution.)

One way to obtain a basic variable for each equation which does not have one
that can easily be picked out[8] is simply to create one. For example, the nonnegative variable (say) X_{a1} is added to equation (4)' and X_{a2} to equality (5)'.

---

[8] For example, there is no variable in the equation which does not appear in any
other one.

This results in

$$-2X_1 + 2X_2 - X_3 \quad + X_{a1} \qquad = 1 \qquad (4)''$$

$$2X_1 + X_2 \quad - X_4 \quad + X_{a2} = 2 \qquad (5)''$$

$$X_1, X_2, X_3, X_4, X_{a1}, X_{a2} \geq 0. \qquad (6)''$$

The trouble, of course, is that the feasible region defined by (4)'-(6)' is not the same as the one given by (4)''-(6)''. However, notice that a feasible solution to (4)''-(6)'' is a feasible solution to (4)'-(6)' provided that $X_{a1} = X_{a2} = 0$. Thus, we could find a basic feasible solution to (4)'-(6)' if a basic feasible solution to (4)''-(6)'' can be produced with $X_{a1} = X_{a2} = 0$. To do this we use the simplex method. In particular, we solve the linear program

$$\text{Maximize} \quad -3X_1 - 4X_2 \qquad - MX_{a1} - MX_{a2} = X_0 \qquad (3)''$$

$$\text{subject to} \quad -2X_1 + 2X_2 - X_3 \quad + X_{a1} \qquad = 1 \qquad (4)''$$

$$2X_1 + X_2 \quad - X_4 \qquad + X_{a2} = 2 \qquad (5)''$$

$$\text{and} \qquad X_1, X_2, X_3, X_4, X_{a1}, X_{a2} \geq 0, \qquad (6)''$$

where M is some very large positive number. Notice that the objective function (3)'' is (3) where the terms $-MX_{a1}$ and $-MX_{a2}$ have been added. The $-M$ coefficient makes it very undersirable to have either X_{a1} or X_{a2} at any positive value since it would make X_0 very small. This means that if any optimal solution to (3)''-(6)'' has X_{a1} and/or X_{a2} positive then the linear program (3)-(6) has no feasible solution. Therefore, subsequent to solving the enlarged problem, an optimal solution is found with either $X_{a1} = X_{a2} = 0$ which means that we have an initial basic feasible solution to the original problem, or some X_{a1} and/or X_{a2} is positive in which case the original linear program has no feasible solution.

To solve the linear program (3)''-(6)'' we pose the objective function (3)'' as a constraint yielding

$$X_0 + 3X_1 + 4X_2 \qquad + MX_{a1} + MX_{a2} = 0 \qquad (3)''$$

$$- 2X_1 + 2X_2 - X_3 \quad + X_{a1} \qquad = 1 \qquad (4)''$$

$$2X_1 + X_2 \quad - X_4 \quad + X_{a2} = 2. \qquad (5)''$$

The variable X_{a1} and X_{a2} has been introduced to serve as a basic variable in equation (3)'' and (4)'', respectively. As each basic variable is only to appear in the corresponding equation, X_{a1} and X_{a2} must be eliminated from (3)''. This can be accomplished by multiplying equalities (4)'' and (5)'' by $-M$ and adding the resulting equation to (3)''. The result of these calculations are exhibited in tableau #1 below, where tableau #0 represents (3)''-(5)'' in detached coefficient form. The remaining tableaux represent the solution to the problem using the standard simplex method. Pivot elements are circled.

Tableaux	Basic variables	X_0	X_1	X_2	X_3	X_4	X_{a1}	X_{a2}	Basic feasible solution
#0	X_0	1	3	4	0	0	M	M	0
	X_{a1}	0	-2	2	-1	0	1	0	1
	X_{a2}	0	2	1	0	-1	0	1	2
#1	X_0	1	3	-3M+4	M	M	0	0	-3M
	X_{a1}	0	-2	②	-1	0	1	0	1
	X_{a2}	0	2	1	0	-1	0	1	2
#2	X_0	1	-3M+7	0	(-1/2)M+2	M	(3/2)M-2	0	(-3/2)M-2
	X_2	0	-1	1	-1/2	0	1/2	0	1/2
	X_{a2}	0	3	0	1/2	-1	-1/2	1	3/2
#3	X_0	1	0	0	5/6	7/3	M-5/6	M-7/3	-11/2
	X_2	0	0	1	-1/3	-1/3	-1/3	-1/3	1
	X_1	0	1	0	1/6	-1/3	-1/6	1/3	1/2

The optimal solution to the enlarged linear program is exhibited in tableau #3. As both X_{a1} and X_{a2} have value 0, we have a basic feasible solution to the original problem or an initial tableau from which to start the simplex algorithm. The tableau is obtained by omitting the columns representing the created variables from the final tableau #3. This produced tableau #4. As all entries in the X_0 row are nonnegative, no more work is required and X_0 = -11/2, X_1 = 1/2 and X_2 = 1 with X_3 = X_4 = 0 is an optimal basic feasible solution.

Tableau	Basic variables	X_0	X_1	X_2	X_3	X_4	Basic feasible solution
#4	X_0	1	0	0	5/6	7/3	-11/2
	X_2	0	0	1	-1/3	-1/3	1
	X_1	0	1	0	1/6	-1/3	1/2

Remarks

1. The variables which are created are called *artificial variables*. These are added only to those constraints which do not immediately exhibit a basic variable. For example, in the system below an artificial variable is added only to expressions (8) and (10). In inequality (7) the slack variable serves as the basic variable, and by dividing equation (9) by 2, X_4 can be designated a basic variable. The resulting system is below where the basic variables are underlined. Notice that a surplus variable must be added to (10) to obtain an equality. Then the artificial variable is adjoined.

$$5X_1 + X_2 \qquad\qquad \leq 4 \qquad (7)$$

$$X_1 + 2X_2 - X_3 \qquad = 3 \qquad (8)$$

$$10X_1 - 4X_2 \qquad + 2X_4 = 6 \qquad (9)$$

$$-3X_2 - 4X_2 \qquad\qquad \geq 5 \qquad (10)$$

$$5X_1 + X_2 \qquad + \underline{X_5} \qquad\qquad = 4 \quad (7)' \text{, slack variable}$$
$$X_5 \text{ is adjoined}$$

$$X_1 + 2X_2 - X_3 \qquad + \underline{X_{a1}} \qquad = 3 \quad (8)' \text{, artificial variable } X_{a1} \text{ is added}$$

$$5X_1 - 2X_2 \quad + \underline{X_4} \qquad\qquad = 3 \quad (9)' \text{, equation (9) divided by 2}$$

$$-3X_1 - 4X_2 \qquad\qquad - X_6 \quad + \underline{X_{a2}} = 5 \quad (10)' \text{, surplus variable } X_6 \text{ and artificial variable } X_{a2} \text{ added.}$$

2. Had some of the coefficients in the X_0 row been negative in tableau #4, the usual simplex algorithm would be applied to solve the original linear program. The enlarged problem is used solely to produce an initial basic feasible solution to the original one or to show that none exists. This part of the computations is called a *Phase I* routine. The solution to the original problem starting with a basic feasible solution produced by the enlarged problem is called a *Phase II* routine. A simplex algorithm which uses artificial variables is sometimes called a *two phase simplex method* (Figure 13). In Example 7, there was only a Phase I. Phase II was not necessary.

coefficient in the corresponding row has value 0, then the variable cannot be
eliminated. However, in this case it turns out ([1], [3]) that the corresponding

Basic variables	X_0	X_1	X_2	X_3	Basic feasible solution		Basic variables	X_0	X_1	X_2	X_3	Basic feasible solution
X_0	1	1	0	2	4		X_0	1	0	0	5	4
X_{a3}	0	-2	0	3	0		X_1	0	1	0	0	0
X_2	0	2	1	0	1	X_{a1} vanishes and is replaced by X_1	X_2	0	0	1	3	1

Tableau at end of Phase I   Tableau for Phase II

FIGURE 13.   ELIMINATING ARTIFICIAL VARIABLES

constraint is implied by the others (i.e., it is *redundant*) and its row may be
dropped (Figure 14).   Thus, artificial variables have a secondary use.

Basic variables	X_0	X_1	X_2	X_3	Basic feasible solution		Basic variables	X_0	X_1	X_2	X_3	Basic feasible solution
X_0	1	1	0	2	4		X_0	1	1	0	2	4
X_{a3}	0	0	0	0	0	Second row dropped, first constraint is redundant	X_2	0	2	1	0	1
X_2	0	2	1	0	1							

Tableau at end of Phase I   Tableau for Phase II

FIGURE 14.   ARTIFICIAL VARIABLE SHOWING A REDUNDANT CONSTRAINT

5. *Duality and the Dual Simplex Method*

Corresponding to any linear program, written as,

P   Maximize $\sum_{j=1}^{n} c_j X_j = X_0$

subject to $\sum_{j=1}^{n} a_{ij} X_j \leq b_o$ $(i=1,\ldots,m)$

and $X_j \geq 0$ $(j=1,\ldots,n)$,

we define the linear program

D   Minimize $\sum_{i=1}^{m} W_i b_i$

subject to $\sum_{i=1}^{m} W_i a_{ij} \geq c_j$ $(j=1,\ldots,n)$

and $W_i \geq 0$ $(i=1,\ldots,m)$,

to be its *dual*. Problem P is called the *primal*. Notice that the costs (c_j) in P become the right hand side in D and the right hand side (b_i) in P become the costs in D. Also, if P has m constraints and n variables then D has n constraints and m variables. That is, for each constraint i in P there is a dual variable W_i in D, and for each variable X_j in P there is a constraint j in D. An illustration is below.

Example (from Example 7)

P   Maximize $-3X_1 - 4X_2 = X_0$

subject to $2X_1 - 2X_2 \leq -1$ (W_1)

$-2X_1 - X_2 \leq -2$ (W_2)

and $X_1, X_2 \geq 0.$

D   Minimize $-W_1 - 2W_2 = W_0$

subject to $2W_1 - 2W_2 \geq -3$

$-2W_1 - W_2 \geq -4$

and $W_1, W_2 \geq 0.$

Remarks

1. In order to define problem P from problem D, the primal must correspond to a linear program with an objective function which is to be maximized and all constraints must be an inequality of the $\geq$ type. As a minimization problem can be converted to a maximization one by changing the sign of the coefficients in the objective function, and any equation can be written as two inequalities[9] any linear program can be posed as problem P.

2. Writing problem D in the format of problem P gives D' and taking the dual of D', assigning the "dual" variable X_j to each constraint in D', gives the primal

---

[9] For example, minimize $3X_1-2X_2+X_3$ becomes maximize $-3X_1+2X_2-X_3$, and $X_1+3X_2=2$ becomes $X_1+3X_2 \leq 2$ and $X_1+3X_2 \geq 2$ or $-X_1-3X_2 \leq -2$.

problem P. Thus, the dual of the dual is the primal.

D'   Maximize $\sum_{i=1}^{n} W_i(-b_i) = (-W_0)$

            subject to $\sum_{i=1}^{m} W_i(-a_{ij}) \leq (-c_j)$ $(j=1,\ldots,n)$ (X_j)

            and $W_i \geq 0$ $(i=1,\ldots,m)$.

3. The dual problem is defined, because it turns out (see [1],[3]) that by
solving problem D we can solve problem P (and vice versa). If problem D is
easier to solve than P, then this result is quite useful. To show how solving
one problem solves theother we use an example and illustrate the *dual simplex
method*.

The Dual Simplex Method

    To illustrate the procedure we solve Example 7, the primal problem, by
solving its dual. For ease of exposition, these programs are rewritten below.

P   Maximize $-3X_1 - 4X_2 = X_0$

            subject to $2X_1 - 2X_2 \leq -1$ (X_{s1})

                       $-2X_1 - X_2 \leq -2$ (X_{s2})

            and $X_1, X_2 \geq 0$.

D   Minimize $-W_1 - 2W_2 = W_0$

            subject to $2W_1 - 2W_2 \geq -3$ (W_{s1})

                       $-2W_1 - W_2 \geq -4$ (W_{s2})

            and $W_1, W_2 \geq 0$.

Adding primal slack variables X_{s1} and X_{s2} to P, and dual slack variables W_{s1} and
W_{s2} to D, we may rewrite P as P' and D as D'. In each problem, the objective
function is to be maximized and is rewritten as a constraint.

P'   Maximize X_0

            subject to $X_0 + 3X_1 + 4X_2 \qquad = 0$

                       $2X_1 - 2X_2 + X_{s1} \quad = -1$

                       $-2X_1 - X_2 \qquad +X_{s2} = -2$

            and $X_1, X_2, X_{s1}, X_{s2} \geq 0$.

D'   Maximize $\bar{W}_0 = (-W_0)$

         subject to $\bar{W}_0 - W_1 - 2W_2 \qquad\qquad = 0$

$$-2W_1 + 2W_2 + W_{s1} \qquad = 3$$

$$2W_1 + W_2 \qquad + W_{s2} = 4$$

         and $W_1,\ W_2,\ W_{s1},\ W_{s2} \geq 0.$

Problems P' and D' are each in canonical form. However, P' does not have a nonnegative right hand side while D' does exhibit a basic *feasible* solution. Thus, using the standard simplex method we may solve D'. The computations appear in tableaux D1 through D3 below. The pivot elements are circled.

Tableau for Dual

Tableau	Basic variables	$\bar{W}_0$	W_1	W_2	W_{s1}	W_{s2}	Basic feasible solution	
				IN				
D1	$\bar{W}_0$	1	-1	--2	0	0	0	
	W_{s1}	0	-2	②	1	0	3	OUT
	W_{s2}	0	2	1	0	1	4	
			IN					
D2	$\bar{W}_0$	1	-3	0	1	0	3	
	W_2	0	-1	1	1/2	0	3/2	
	W_{s2}	0	③	0	-1/2	1	5/2	OUT
D3	$\bar{W}_0$	1	0	0	1/2	1	11/2	
	W_2	0	0	1	1/3	1/3	7/3	
	W_1	0	1	0	-1/6	1/3	5/6	

Problem D can also be solved starting with tableau P1 for problem P'. In particular, we may correspond tableau P1 with D1 by making the following primal-dual variable correspondence.

Primal . . . objective function	variables		slack variables		
X_0	X_1	X_2	X_{s1}	X_{s2}	
$-\bar{W}_0 = W_0$	W_{s1}	W_{s2}	W_1	W_2	(1)
Dual . . . objective function	slack variables		variables		

H. SALKIN AND J. SAHA

That is, primal variable X_j is associated with the dual slack variable W_{sj} ($j=1,\ldots,n$), and the primal slack variable X_{si} corresponds to the dual variable W_i ($i=1,\ldots,m$). By doing this we can pick out the values of the dual basic feasible solution in tableau D1 from the X_0 row in tableau P1. In particular, from the X_0 row: $-\overline{W}_0 = 0$, $W_{s1} = 3$, $W_{s2} = 4$, and $W_1 = W_2 = 0$. Also, notice that X_{s1} and X_{s2} are basic variables in tableau P1, and using the association (1) we have that W_1 and W_2 are nonbasic variables in the current basic feasible solution to D'. This agrees with tableau D1.

The entering or IN variable, W_2, in tableau D1 corresponds to the most negative coefficient in the $\overline{W}_0$ row. In tableau P1 this is equivalent to the largest negative value for the primal basic variable (excluding the objective function), namely X_{s2}. Thus, suppose we designate X_{s2} as the OUT variable in tableau P1.

Tableaux for Primal

Tableau	Basic variables	X_0	X_1	X_2	X_{s1}	X_{s2}	Basic feasible solution
			IN				
P1	X_0	1	3	4	0	0	0
	X_{s1}	0	2	-2	1	0	-1
	X_{s2}	0	(-2)	-1	0	1	-2 OUT
				IN			
P2	X_0	1	0	5/2	0	3/2	-3
	X_{s1}	0	0	(-3)	1	1	-3 OUT
	X_1	0	1	1/2	0	-1/2	1
P3	X_0	1	0	0	5/6	7/3	-11/2
	X_2	0	0	1	-1/3	-1/3	1
	X_1	0	1	0	1/6	-1/3	1/2

To find the basic variable to leave the basis, or the OUT variable, in
tableau D1, we scanned the column corresponding to W_2 for the positive values
which are the coefficients 2 and 1. Then the ratios 3/2 and 4/1 were formed,
and the minimum was found. This resulted in W_{s1} as the OUT variable. In tableau
P1 the values 2 and 1, are in the *row* corresponding to the OUT variable X_{s2}, and
appear as -2 and -1, respectively. Moreover, the right hand side values 3 and 4
in tableau D1 appear as 3 and 4 in the X_0 row of tableau P1. To obtain the
equivalant ratios and pivot element in tableau P1 as in tableau D1, we simply
choose the variable corresponding to the minimum of $3/|-4|$ and $4/|-1|$ as the IN
variable; or X_1 is the IN variable. The pivot element in tableau P1 is thus -2.

At this point we have that the basic variable X_{s2} in tableau P1 is to be
replaced by the nonbasic variable X_1. Using the association (1) this means that
the nonbasic variable W_2 is replacing the basic variable W_{s1}. This agrees with
tableau D1.

Performing a standard pivot operation produces tableau P2. Notice that again
using (1) we can find the basic feasible solution to the dual linear program
appearing in tableau D2 from the X_0 row in tableau P2. Moreover, using our previous analysis on tableau P2, X_{s1} is the OUT variable and X_2 is the IN variable.
The pivot element is -3. Subsequent to a pivot we have the optimal tableau and
the primal problem has been solved.

Notice that all we have done is solved the dual problem D' using a tableau
for its primal problem P'. The basic feasible solutions to the dual problem
appear in the X_0 row of the tableau for the primal (see Table 1). By solving the
dual problem using tableaux for the primal, we have also solved the primal linear
program. It turns out that if one problem (i.e., primal or dual) has an optimal
solution, then so does the other and this tableaux correspondance can always be
made. The optimal tableau for D will also correspond to the optimal tableau for
P (and vice versa).

Tableau	Primal		Dual	
	Basic variables	Nonbasic variables	Basic variables	Nonbasic variables
P1	$X_{s1}=-1$, $X_{s2}=-2$	$X_1=X_2=0$	$W_{s1}=3$, $W_{s2}=4$	$W_1=W_2=0$
P2	$X_{s1}=-3$, $X_1=1$	$X_2=X_{s2}=0$	$W_2=3/2$, $W_{s2}=5/2$	$W_1=W_{s1}=0$
P3	$X_1=1/2$, $X_2=1$	$X_{s1}=X_{s2}=0$	$W_1=5/6$, $W_2=7/3$	$W_{s1}=W_{s2}=0$

TABLE 1.   PAIRS OF BASIC SOLUTIONS

If the rules for computing the OUT and IN variable in the primal tableaux are remembered, then the dual problem can always be solved without it ever explicitly appearing. In particular, tableaux D1 through D3 need never appear and only tableaux P1 through P3 are needed.

Finally, notice that the original linear program corresponded to a dual problem which initially displayed a basic *feasible* solution. This is equivalent to having nonnegative values in the X_0 row (see tableau P1). When this occurs, a Phase I routine or artificial variables are not needed, and we can immediately proceed to solve the dual problem and hence the original one. Therefore, using the technique described here, called the *dual simplex method*, Example 7 can be solved directly via tableaux P1 through P3 rather than tableaux #0 through #3 (p. 24) which contained artificial variables. The various possible initial tableaux and suggested solution technique appear in Figure 15. The advantages of the primal or standard simplex method versus the dual simplex method are given in Table 2.

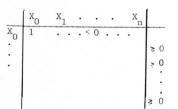

(a) Primal simplex method
     (without artificial
     variables)

(b) Dual simplex method

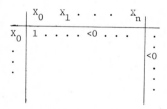

(c) Primal simplex method
     (with artificial variables)

FIGURE 15.   POSSIBLE SITUATIONS FOR THE INITIAL TABLEAU

Technique	Primal	Dual
Advantage	A basic feasible solution is produced at each iteration.	A Phase I routine is not necessary.
Disadvantage	Artificial variables or a Phase I routine may be needed.	A feasible solution is never produced until an optimal one is found.

TABLE 2.

Listing of the Dual Simplex Algorithm

Using the general tableau below, we can list the technique.

Basic variables	X_0	X_1 . . . . X_n	Basic solution
X_0	1	a_{01} . . . a_{0n}	b_0
X_{B1}	0	a_{11} . . . a_{1n}	b_1
.	.	. .	.
.	.	. .	.
.	.	. .	.
X_{Bm}	0	a_{m1} . . . a_{mn}	b_m

Step 1. Place the original problem in canonical form so that $a_{01} \geq 0, \ldots, a_{0n} \geq 0$. Go to step 2.

Step 2. Determine the OUT variable X_{Br} using $b_r = \text{minimum}_i\, b_i$ $(i \geq 1)$, such that $b_i < 0$. If $b_i \geq 0$ for $i=1,\ldots,m$, the current basic solution is feasible and optimal - terminate. Go to step 3.

Step 3. Find the IN variable X_k by $\dfrac{a_{0k}}{|a_{rk}|} = \text{minimum}_j \dfrac{a_{0j}}{|a_{rj}|}$, where $a_{rj} < 0$. If $a_{rj} \geq 0$ for $j=1,\ldots,n$, the problem has no finite optimal solution[10] - terminate. Go to step 4.

Step 4. Using a_{rk} as the pivot element, perform a pivot operation and go to step 2.

---

[10] In this case, the dual has an unbounded solution and the primal or original problem either has no feasible solution or an unbounded solution (see [1], or [3]).

Economic Interpretation of the Dual Variables

Consider the following problem:

$$\text{Maximize} \quad 5X_1 + 3X_2$$

$$\text{subject to} \quad 3X_1 + 5X_2 \leq 15 \quad (1)$$

$$5X_1 + 2X_2 \leq 10 \quad (2)$$

$$\text{and} \quad X_1, X_2 \geq 0.$$

Let $(\hat{X}_1, \hat{X}_2)$ be an optimal solution to this problem. The dual problem is:

$$\text{Minimize} \quad 15W_1 + 10W_2$$

$$\text{subject to} \quad 3W_1 + 5W_2 \geq 5 \quad (3)$$

$$5W_1 + 2W_2 \geq 3 \quad (4)$$

$$\text{and} \quad W_1, W_2 \geq 0.$$

Let $(\hat{W}_1, \hat{W}_2)$ be an optimal solution of the dual.

If the right hand side value in the constraint (1), which denotes the availability of a resource, is increased to $15+\Delta$, the dual objective function becomes $(15+\Delta)W_1 + 10W_2$. The dual constraints remain unchanged. Suppose this increase does not change the optimal dual solution, i.e., $(\hat{W}_1, \hat{W}_2)$ is still an optimal solution of the dual. Let $(\hat{\hat{X}}_1, \hat{\hat{X}}_2)$ be an optimal solution of the new primal problem. Using the fact (see [1] or [3]) that at any pair of optimal solutions, the value of each objective function is equal (called the *dual theorem*), we have

$$5\hat{\hat{X}}_1 + 3\hat{\hat{X}}_2 = (15+\Delta)\hat{W}_1 + 10\hat{W}_2. \quad (5)$$

Also, since $(\hat{X}_1, \hat{X}_2)$ and $(\hat{W}_1, \hat{W}_2)$ are optimal solutions of the original primal and the dual, again by the dual theorem, we have

$$5\hat{X}_1 + 3\hat{X}_2 = 15\hat{W}_1 + 10\hat{W}_2. \quad (6)$$

Thus, using (5) and (6), we have

$$5\hat{\hat{X}}_1 + 3\hat{\hat{X}}_2 = (15+\Delta)\hat{W}_1 + 10\hat{W}_2$$

$$= (15\hat{W}_1 + 10\hat{W}_2) + \Delta\hat{W}_1$$

$$= (5\hat{X}_1 + 3\hat{X}_2) + \Delta\hat{W}_1$$

Or, solving for $\hat{W}_1$, yields

$$\hat{W}_1 = \frac{(5\hat{\hat{X}}_1 + 3\hat{\hat{X}}_2) - (5\hat{X}_1 + 3\hat{X}_2)}{\Delta}.$$

This shows that $\hat{W}_1$, the optimal value for the first dual variable, is the rate of change of the maximum value of the primal objective function with respect to the unit change in the right hand side value of the constraint (1). We can show similar economic interpretation for W_2. Therefore, in this sense, the dual variable W_i, can be interpreted as a measure of the rate of change in the value of the primal objective function per unit change in the right hand side element b_i.

6. *Computer Implementation and Codes*

In this section, we discuss some of the finer points of linear programming algorithms in context of computer implementation. Then a listing and brief discussion of several commercially available codes are given. We are indebted to the individuals and corporations listed in the appendix for their time and assistance.

The Basis Inverse

We have described the primal and dual simplex methods in a very simple way. However, they could have been presented in a more sophisticated manner using matrix algebra and certain results from linear algebra. If this had been done, it would be clear that for computer implementation only certain information is needed at each iteration in the simplex computations rather than the entire simplex tableau. In particular, if there are m *nonredundant* constraints[11] and n variables then, besides the original data, only the *inverse* of the *basis matrix*[12] and a list of which variables are basic ones need be kept. Here, the basis matrix is the square array formed by taking the original coefficient columns corresponding to the basic variables. For example, from tableau P2 in the last illustration, the basic variables are X_0, X_{s1} and X_1, so the basis matrix B is

$\begin{pmatrix} 1 & 0 & 3 \\ 0 & 1 & 2 \\ 0 & 0 & -2 \end{pmatrix}$. The inverse of B, written B^{-1}, can be computed to give $\begin{pmatrix} 1 & 0 & 3/2 \\ 0 & 1 & 1 \\ 0 & 0 & -1/2 \end{pmatrix}$.

Once B^{-1} is computed, all other values that normally appear in the simplex tableau can be generated.

---

[11] That is, m constraints where no one is implied by the others.

[12] If B is a square matrix, where none of its columns can be expressed as a linear combination of the others, then its inverse B^{-1} exists, and using matrix multiplication $BB^{-1}=B^{-1}B=I$ (an identity matrix).

As the basis inverse may be substantially large several well known techniques have been developed to reduce storage requirements. Most notably these include keeping the inverse in *product form* or in *lower-upper triangular (i.e., "LU") form*. These procedures are common to most linear programming computer systems and a description may be found in [5]. Also, to reduce numerical roundoff errors and overflow problems, these systems contain subroutines which periodically reinvert the basis. That is, rather than continuously find the basis inverse from the previous one,[13] at some iteration, B^{-1} is recomputed starting with the original coefficient columns. Naturally, a *reinversion routine* may consume a substantial amount of computation time.

The Primal-Dual Simplex Methods

Besides the primal and dual simplex methods, *primal-dual simplex algorithms* have been developed. These techniques are used when neither the primal nor dual problem initially exhibits a basic feasible solution and a Phase I routine is not desired (see Figure 15(c)). The procedure, starting with a tableau as in Figure 15(c), solves the primal and dual problem simultaneously. A few computer systems contain algorithms of this type. The details of the technique are left to the references.

Decomposition and Generalized Upper Bounding (GUB)

Most linear programs contain coefficient matrices which are highly structured. In particular, some problems may appear as in Figure 16(a) and/or contain constraints of the type given in Figure 16(b). Using the simplex method, a basis of size m by m is kept and updated; where m is the number of constraints. However, if a problem appears as in Figure 16(a), it turns out ([1],[3]) that it can be solved using a series of small inverses corresponding to matrices taken from each (matrix) block $A_1,\dots,A_t$. In many instances this can result in an enormous reduction of computer storage requirements. However, the technique, termed the *decomposition method*, does require more computations at each iteration than the normal simplex method and thus may result in more running time. As of this writing, the authors are not aware of any commercially available code containing a decomposition method.

Constraints of the type appearing in Figure 16(b) are termed *multiple choice constraints*. They bound each variable, as well as a sum of variables, by one.

---

[13] The details of these computations are in [1] and [3].

A technique, termed *generalized upper bounding* (GUB) (see [5]) has been developed
which allows these to be dropped from the linear program. These constraints are
implicitly kept and are explicitly enforced at each iteration. Naturally, the
user must input these bounding constraints in some form. Moreover, simple bounding constraints (e.g., $X_j \leq 3.7$) also need never explicitly appear and special
techniques have been developed for these as well. Some of the more recent computer codes contain a GUB capability, and nearly all have routines to take simple
upper bounds into account.

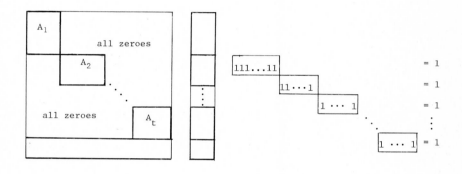

(a) Decomposable Problem   (b) Multiple Choice Constraints

FIGURE 16. SPECIAL SITUATIONS

Parametric Programming

    In parametric programming we study the behavior of solutions to a linear
programming problem when the data for the problem are allowed to vary. For
example, we could vary the cost coefficients, the coefficients in the constraints,
or the right hand side values. This analysis is important since the numerical
values of various coefficients may not be accurate.
    Some work has been done with varying cost coefficients and the right hand
side values. The details are left to the references ([1],[3]). The study of
the behavior of solutions with variations in the elements in the constraint
matrix is complex. A detailed account may be found in Gass [2].

Separable Programming

In many problems the objective function is not linear but additively separable, i.e.,

$$Z = \sum_{j=1}^{n} f_j(X_j);$$

where, $f_j(X_j)$ is a nonlinear function and Z is to be maximized or minimized. There are methods for linearizing these functions which convert the objective function into a linear form. A detailed exposition of separable programming, which studies these linearization techniques, may be found in Dantzig [1].

Commercially Available Codes

The first linear programming problem was solved on a digital computer in January 1953, with a National Bureau of Standards computer, using SEAC ([2]). Since that time, computer technology has vastly improved and various computer codes have been developed for solving large scale linear programming problems. Today it is possible to solve a problem with as many as 50,000 constraints and 285,000 variables in a relatively reasonable amount of time. We now list and describe several of the contemporary codes.

1. The Univac Code

Name   . . . UNIVAC's Mathematical Programming (MP) System -- FMPS.
Source   . . . Local UNIVAC branch office.
Language   . . . UNIVAC 1100 series assembly language and UNIVAC FORTRAN V
                    language.
Limitations   . . . Maximum of 8,192 rows in the coefficient matrix, dependent on
                    the main storage availability.
Cost   . . . It is made available free of charge to UNIVAC customers.
Computations . . . UNIVAC claims that for most problems it is the fastest commer-
                    cial mathematical programming system available.

The system includes a zero-one mixed integer program branch and bound algorithm, separable programming, GUB, parametric programming and procedures which can be called during the solution process to perform nonlinear adjustments of the matrix coefficients. It also includes a FORTRAN-like command language, and a generalized matrix generator/report writer called Gamma 3.

2. IBM Code

Name   . . . Mathematical Programming System Extended (MPSX) with mixed
                     integer programming (MIP) and generalized upper bounding (GUB)
                     feature; program product number 5734-XM4.

Source   . . . IBM sales representative at any local IBM office.

System   . . . The minimum configuration required by MPSX, which operates
                     under OS or OS/VS, is an IBM System/360 or System/370 (128k)
                     with decimal and floating point instruction sets. The minimum
                     region size is 86k, and direct access storage must also be
                     available.

Language   . . . System/360 assembly language.

Limitations   . . . Logical processing capabilities of 16,383 rows for LP, 65,535
                     rows for GUB (with a secondary limit of not more than 16,383
                     rows for the LP part), and 4,095 integer variables for the MIP.

Computations   . . . As of September 1974 the program has been in use for three and
                     one half years.

Cost   . . . Monthly licensing fees are $132 for the basic package, plus an
                     additional $229 for MIP and $663 for GUB capabilities. Both
                     basic LP and GUB include separable programming capabilities.

Manuals   . . . General Information   #GH20-0849
                     Introduction to Modeling Using MIP   #GE19-5043
                     Control Language Users Manual   #SH20-0932
                     Linear and Separable Programming -
                     Program Description Manual   #SH20-0968
                     MIP Feature - Program Description Manual   #SH20-0908

   MPSX contains all the capabilities and features that are available with
MPS/360, such as a control program containing its own language and a compiler; a
set of linear and separable programming procedures; a FORTRAN interface program
called Read Communications Format; access to the MARVEL program, which contains
a language processor for matrix generation, output analysis, and report writing;
and an interface to the Report Generator Program (MPSRG).

3. IBM Code

Name   . . . MPSX/370. Program Product Numbers 5740-XM3 (OS/VS version) and
                     5746-XM2 (DOS/VS version). Each version has a mixed integer
                     programming (MIP) algorithm.

Source   . . . IBM sales representative at any local IBM branch office.

System   . . . System/370 running under OS/VS or OS/DOS.

Language   . . . System/370 assembly language.

Limitations   . . . Logical processing capabilities of 16,383 rows for LP, and
                    32,767 integer variables.

Computations . . . Limited to test sites use until the code is released during
                    January 1975.

Cost   . . . The monthly licensing fees are as follows:

MPSX/370	OS/VS	$400
MIP/370	OS/VS	$350
MPSX/370	DOS/VS	$350
MIP/370	DOS/VS	$300

Manuals   . . . General Information Manual   #GH19-1090
                    Primer   #GH19-1091

The basic LP system includes separable programming. MPSX/360 contains all
the special features of MPSX.

4. SCICON Code

Name   . . . UMPIRE

Source   . . . Scientific Control Systems Ltd.
                    Sanderson House
                    49-57 Berners Street
                    London W1P 4AQ

System   . . . UNIVAC 1106, 1108, 1110 under Exec. 8.

Language   . . . FORTRAN V, UNIVAC, 1100 assembler.

Limitations   . . . 8000 rows, 256,000 columns, and core requirements 30k upwards.

Computations . . . Not available.

Cost   . . . Computer time, approximately: 90 rows, 120 columns, $12.50
                    540 rows, 1,060 columns, $200.

Comprehensive User's Guide available from Scientific Control Systems Ltd.
UMPIRE may be used outside the United Kingdom. The system contains a GUB algorithm.

5. NBER Code

Name   . . . SESAME

Source   . . . National Bureau of Economic Research, Inc.
                    575 Technology Square
                    Cambridge, Massachusetts 02139

System   . . . SESAME now operates under the Cambridge Monitor System (CMS) on
                    the IBM System/360, Model 67 computer.

Language   . . . OS/360 assembler language.

Limitations . . . Not available.

Cost   . . . Not available.

SESAME is controlled by its own interactive executive routine, and a limited control language. It will be transferred to operate under ACOS/ACOL, which is a comprehensive application-control operation system/language now under development at the NBER Computer Research Center. ACOS/ACOL will run under CP-67 on the System/360, Model 67 computer, and under OS/VS II on the System/370, Models 158 and 168 computers. SESAME is unique in that it represents an interactive mathematical programming system.

6. *Honeywell Code*

Name   . . . LP 6000 System.

Source   . . . Systems representative at any local Honeywell branch office.

System   . . . Honeywell Series 6000 or 600 central system and console with
                   35k words of main memory, exclusive of memory required for the
                   operating system.
                   - Magnetic tape subsystem with at least one tape unit.
                   - Bulk storage (disk or drum).
                   - Card reader.
                   - Printer.
                   - Card Punch (if punched card output is required).

Language   . . . ?

Limitations . . . 4,095 rows and 262,000 columns for standard problems; a trans-
                   portation matrix may have more than 20,000 rows and columns,
                   consisting of up to 4,095 sources and 16,000 destinations, or
                   vice versa.

Cost   . . . Contact any Honeywell local branch office.

The LP 6000 System contains the Agenda Control Language (ACL) for calling and executing the algorithms and other programs. The system has the Matrix Generator Language (MGL) which is used to convert and store the user's problem data in the format compatible with the input requirements of the LP 6000 System. The system also includes the Format Generator Language to control and define the output reports from the LP 6000 System. The LP 6000 System has separate primal, dual, integer, transportation, and several other solution algorithms.

7. CDC Codes

Name   . . . OPHELIE: A Linear Programming Code

Source   . . . SIA International
                    35 Boulevard Brune
                    Paris 14e, France

System   . . . CDC 6600, requirements either 64k or 128k bytes.

Language   . . . ?

Limitation   . . . 20,000 rows

Cost   . . . $45,000

The package contains a mixed integer programming routine (OPHGEMIK), a matrix generator code, a report write routine (OPHEDIT), and a linear programming core program (OPHELIE).

Name   . . . OPHELIE II

Source   . . . Local branch office of Control Data Corporation.

System   . . . Control Data 6000 and Cyber 70 series computer systems.

Language   . . . ?

Limitation   . . . ?

Cost   . . . ?

The package contains three subsystems. OPHELIE/LP is the core of OPHELIE II. OPHELIE/MGL is the matrix generator subsystem which allows the user to transform data to the format acceptable to the OPHELIE II system. The report generator subsystem, OPHELIE/RGL, is a convenient language for presenting the output from OPHELIE II in a meaningful fashion.

8. Bonner and Moore Code

Name   . . . Functional Mathematical Programming System (FMPS)

Source   . . . Monte G. Smith, President
                    Bonner and Moore Software Systems
                    500 Jefferson Street
                    Houston, Texas 77002

System   . . . The package is designed for IBM 360/370 series and UNIVAC 1100
                    series computers.

Language   . . . FORTRAN and ASSEMBLER.

Limitation   . . . ?

Cost   . . . Available on request.

FMPS includes a FORTRAN-like programming control language. It has FORTRAN compatible files for input. The FMPS matrix generator and report writer language is Gamma 3. The program includes linear programming, separable programming, and mixed integer programming algorithms with upper bounding features.

9. *Management Science System Code*

Name   . . . MPS III

Source   . . . Charles M. Boudrye, Vice President
                    Management Science Systems
                    121 Congressional Lane
                    Rockville, Maryland 20852

System   . . . IBM 360/370, under OS, PCP-BAL, or FORTRAN.

Language   . . . ?

Limitations   . . . ?

Cost   . . . Basic $200/month

                    GUB $750/month

                    DATAFORM $800/month

MPS III utilizes the IBM standard MPS formats for input, output, problem
files, and control language. Additional features include a GUB capability, a data
management language – DATAFORM, MISTIC for solving MIP, and SAGE – a set of
procedures for reducing processing time.

10. *Haverly Systems Code*

Name   . . . LP/360

Source   . . . G.M. Lowell
                    Haverly Systems, Inc.
                    4 Second Avenue
                    Denville, New Jersey 07838

System   . . . The program is designed for IBM 360 series computer systems.

Language   . . . ?

Limitations   . . . Core memory   Number of rows

                        32k   300

                        64k   1000

                       265k   4000

Cost   . . . $7,500 - $25,000 or leases available.

This is a comprehensive linear and mathematical programming system. It has
the following features: linear optimization, separable optimization, and mixed
integer optimization. MAGEN IV is the general purpose matrix generator language
and report writer included in the FMPS system.

11. *NCR Code*

Name   . . . LP 16

Source   . . . Local NCR branch office.

System   . . . NCR Century, 16k - Neat/3.

Language   . . . ?
Limitations . . . 50 rows and 180 columns.
Computations . . . ?
Cost   . . . Negotiable.

12. *NCR Codes*

Name   . . . LPS, NCR Century Linear Programming System
Source   . . . Local NCR branch office.
System   . . . NCR Century System with at least 32k of memory.
Language   . . . ?
Limitations . . . 150 rows and 1000 columns with a maximum of 6,150 nonzero
                     elements.
Computations . . . Not available.
Cost   . . . Negotiable.

Name   . . . Multifile Linear Programming – MFLP
Source   . . . Local NCR branch office.
System   . . . NCR 315, 10k – NEAT. The package consists of four programs
                     which are written in modular form. LPI is similar to a matrix
                     generator. LPC is an updating program. LP solves and stores
                     the result on a magnetic media, and LPO is for sensitivity
                     analysis.
Language   . . . ?
Limitations . . . ?
Cost   . . . Negotiable.

References

1. Dantzig, G.B.; Linear Programming and Extensions, Princeton University Press,
        Princeton, New Jersey (1963).

2. Gass, S.F.; Linear Programming Methods and Applications, McGraw-Hill Book
        Company, New York, New York (1958).

3. Hadley, G.; Linear Programming, Addison-Wesley Publishing Company, Reading,
        Mass. (1963).

4. ICP Quarterly, ed. L.A. Welke, International Computer Program, 2506 Willow-
        brook Parkway, Indianapolis, Indiana 46205.

5. Lasdon, L.S.; Optimization Theory for Large Systems, The Macmillan Company,
        New York, New York (1970).

6. Salkin, H.S.; Integer Programming, Addison-Wesley Publishing Company, Reading,
        Mass. (1975).

7. SESAME: Design and Capabilities Overview, Computer Research Center for
             Economics and Management Science, National Bureau of Economic
             Research, Inc., 575 Technology Square, Cambridge, Mass.   02139
             (1964).

8. Thomas, G.B.; <u>Calculus and Analytical Geometry</u>, Addison-Wesley Publishing
             Company, Reading, Mass. (1962).

Appendix: Acknowledgments for Section 6

1. Dr. E. Martin Beale
    Dr. P. Bryan Tatham
    Scientific Control Systems Limited
    Sanderson House
    49-57 Berners Street
    London, W1P 4AQ

2. Dr. Leon S. Lasdon
    Department of Operations Research
    Case Western Reserve University
    Cleveland, Ohio 44106

3. Dr. E.H. McCall
    UNIVAC Division
    P.O. Box 3942
    St. Paul, Minnesota

4. Mr. Nicholas A. Molley
    Honeywell Information Systems Inc.
    1001 East 55 Street
    Cleveland, Ohio 44103

5. Dr. H.V. Smith
    IBM Corporation
    1133 Westchester Avenue
    White Plains, New York 10604

6. Mr. V.V. Sundaram
    SOHIO
    Midland Building
    Cleveland, Ohio 44114

7. Dr. William W. White
    IBM Corporation
    2651 Strang Boulevard
    Yorktown Heights, New York 10598

PART I

APPLICATIONS FROM INDUSTRIAL ENVIRONMENTS

Chapter 2

A CORPORATE SHORT TERM PLANNING MODEL:

A CASE STUDY IN THE NON-FERROUS METAL INDUSTRY

Bernard A. Lietaer
Cresap, McCormic, and Paget, Inc.
New York, New York 10017

Abstract.   In contrast with many typical linear
programming applications which address themselves
to special functional problems (e.g., transportation, machine loading, and blend mixing), this
chapter describes the development of a corporate
short term planning model. Such a model covers
an important range of functions of the management
process from raw material acquisition planning
to production and marketing. The general concepts underlying short term planning models along
with the industries where these concepts can be
applied are described. This involves the *process
industries* including meat packing, shoe manufacturing, and non-ferrous metals.

The latter part of the chapter is devoted to
a case study in the non-ferrous metals industry;
in particular, a large corporate model developed
in 1971-1972 for the Cerro Corporation in Peru is
developed. This model included mining and concentrating of non-ferrous metal ores, purchase and
sale of concentrates, smelting and refining of the
dozen metals produced by the company, and transportation and marketing of the end products in the
world markets. A submodel, limited to the zinc
circuit, is analysed in more depth, with particular
emphasis on its design, implementation, and usage.
This model has been operating successfully and
extensively tested over one year. Significant
profit improvements can be traced back to this new
system.

1. *Introduction*

This chapter reflects the development, implementation, and testing of a short
term planning model for the Cerro de Pasco Corporation, Peru's largest non-ferrous
metals mining and smelting company. Three aspects made this project an unusually
interesting one. First, the substantial improvement the model has made on the
operating and financial results of the corporation, second, the positive impact
it has had on the balance of payments and usage of national resources in Peru,
and third, the scope of the model itself. We now elaborate on these parts.

Evaluating the financial impact of a model is always difficult. An attempt
at obtaining such a measurement was made with two subparts of the model using
historic data. The first test involved the zinc circuit subpart which will be
described in a subsequent section. The second test concerned the copper-lead
circuit subpart. Both tests covered a recent three month time period. By comparing the actual results with the optimal results which could have been obtained
had the short term planning derived from the model been implemented, an annual additional before tax profit of 1.8 million dollars for the zinc test, over 3 **million**
dollars for the copper-lead test was derived. Since then, the zinc model has been

extensively used with substantial success.

The second aspect is also an important one. In times when host countries are often in conflict of interest with foreign corporations operating in its territory, the model highlights a point of genuine mutual interest. This is especially true in this case where the vast majority of all sales are for export, and thus the model automatically optimizes the hard currency proceeds for the Peruvian resources.

The reason for the operating impact described in the first point can be partially attributed to the third aspect of this study, i.e., the scope of the model. Linear programming models have been successfully applied in short term planning areas such as machine loading, blend mixing for animal feed or gasoline, and transportation scheduling. In contrast with these relatively narrowly defined activities, the scope of the Cerro de Pasco model includes a wide range of management functions from planning the purchase of the raw material, the blending and production process, to marketing of the end products. In addition, the processes which were modeled are relatively complex. They include mining, the concentrating, smelting and refining of a dozen different metals, the production of a series of chemical byproducts, the purchase and sale of concentrates, and the worldwide shipment of the end products. The inclusion of the total managerial process has produced one of the most significant sources of improvements in the operations of the new system. By formally evaluating the impact of one decision on other parts of the corporation, the implementation of the model brought to light significant savings potentials. For example, a new grinder for a concentrating plant costing $60,000, allowed additional profits in the zinc refineries department of over $300,000 per year.

2. *Short Term Planning*

Short term planning is concerned with management desicisons in utilizing the existing and known resources of a company. In general, it involves the planning of the way available raw materials, equipment, manpower, products, and markets should be used to achieve the objective of the corporation. Although the term suggests that the time period considered should be short, there is usually no specified limit. Typically the time involved would be between a few months to a year. However, depending on the industry, a short term plan could involve a time period of a shorter or longer duration. For example, modification of a production plan due to a few days strike, as well as the evaluation of a new raw material sale contract involving three to five years, could be considered as a short term planning problem. Generally, the time period associated with short term planning should be such that all the economic and operating consequences of any decision can be predicted with sufficient precision.

Usually short term planning models are constructed to solve routine operational problems so as to improve on the existing operation. However, a short term planning model could be used to study the implications of specific decisions. A short term planning model aims at optimizing the gross contribution, defined as the difference between the revenue derived from the sales of all end products and all the variable costs incurred to produce them. Variable costs are the ones which change with the level of the production activity. These include the costs of raw materials, labor, transportation, etc. Fixed costs, like fixed overheads and fixed labor expenses, are not included in the short term planning model because they do not depend on any decision incorporated in the model, and will be incurred irrespective of any management decision (except closing down the operation). The decision variables in a short term planning model are the production levels of various end products, called activities. They are usually constrained by the input-output relationships, the available resources, and the policy decisions.

Short term planning models are applicable in a wide variety of industries. However, they are more relevant in industries which, with a given manpower and production facility, have some freedom to choose their raw materials, processing systems, and products or markets. The greater the degree of freedom the higher degree of impact the model could afford. The potential industries include mining, chemicals, pharmaceuticals, textiles, shoes, plastics, meat packing, mechanical equipment, paper, and electronic assembly. In all these industries, it is possible to change inputs, and/or products, and/or production process with a given plant and personnel scenario. In contrast, a short term planning model would not be useful, for example, in a plate glass manufacturing situation which can accept one quality of raw material and whose only degree of freedom is to produce or not to produce a single product.

3. *Short Term Planning Model for the Cerro de Pasco Corporation* (Peru)

History of the Project

The short term planning model for Cerro de Pasco was developed during late 1971 and early 1972, within the framework of a larger project by the Manhattan based consulting firm of Cresap, McCormic and Paget, Inc. This study was directed by Arthur D. Gimbel, aimed at installing a flexible budgeting system at Cerro Corporation, both in the United States and Peruvian operations. During this project, the idea developed between Mr. Gimbel and some members of Cerro de Pasco management, particularly Mr. B. Wadia, that an optimization model of the Peruvian operations could be feasible and worthwhile.

The author was then invited to direct the development and installation of such a model in Peru. A study team was organized which at various times included four analysts, three metallurgists, and seven engineers, familiar with the intricate operations of the different plants. A pilot project, involving the zinc circuits, was undertaken first. A model was developed, tested, and implemented. On successful completion of this project, the model was generalized to include the remaining operations in Peru with its implementation scheduled in different phases over several years. A permanent information updating system was then installed and staff structure was developed.

Cerro de Pasco Operation

Cerro de Pasco, the largest corporation in Peru with over 15,000 employees, extracts, smelts, and refines not only copper, zinc, and lead, but also bismuth, silver, gold, cadmium, selenium, tellivium, indium, and antimony. Although many non-ferrous metal producers have larger tonnage output than Cerro, the unusually large number of ore types extracted and metals treated, and the geographical spread of the operations make this corporation the most complex metallurgical processing plant in the world. The system includes six mining areas located in the remote reaches of the Andes (some operations are at over 15,000 feet). The five million tons of mostly low grade sulphide ores extracted per year are first concentrated and separated into lead, copper, and zinc concentrates for further processing in the smelters. The process is carried out in six concentrating plants, where ores are successively crushed, screened, ground, floated, thickened, and filtrated.

The smelting and refining complex in La Oroya, at over 12,000 feet represents the next process. The products from the six concentrating plants, added to over 100 types of ores and concentrates purchased outside the company's iron plants, are centralized and fed through three different but interconnected circuits, namely, zinc, copper, and lead. Figure 1 schematically represents the flows of materials from the mines to concentrators, and through the three circuits ending in the products shown on the extreme right hand side of the figure. The company has 2 railroads, 4 hydroelectric plants, 35 primary schools, 31 social service centers and housing groups, 9 hospitals, and a livestock ranching operation.

4. Development of the Model

The general model for Cerro de Pasco is, in fact, a grouping of five interconnected submodels. Each of the submodels is a short term planning model. Figure 2 shows the relationship between the five systems. In this chapter

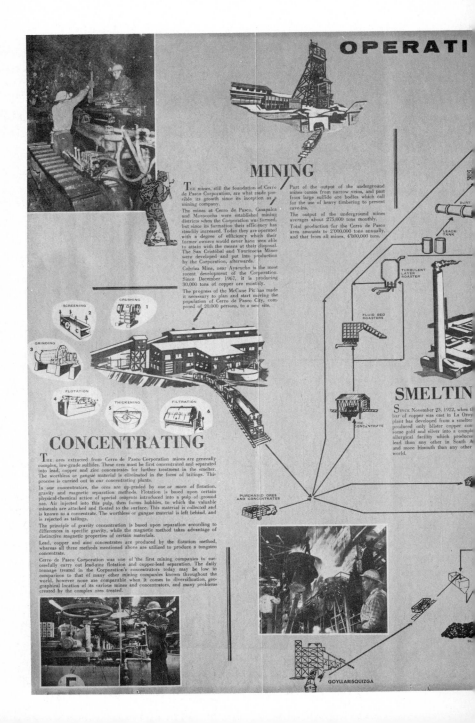

OPERATI

MINING

The mines, still the foundation of Cerro de Pasco Corporation, are what made possible its growth since its inception as a mining company.

The mines at Cerro de Pasco, Casapalca and Morococha were established mining districts when the Corporation was formed, but since its formation their efficiency has steadily increased. Today they are operated with a degree of efficiency which their former owners would never have been able to attain with the means at their disposal. The San Cristóbal and Yauricocha Mines were developed and put into production by the Corporation, afterwards.

Cobriza Mine, near Ayacucho is the most recent development of the Corporation. Since December 1967, it is producing 30,000 tons of copper ore monthly.

The progress of the McCune Pit has made it necessary to plan and start moving the population of Cerro de Pasco City, composed of 20,000 persons, to a new site.

Part of the output of the underground mines comes from narrow veins, and part from large sulfide ore bodies which call for the use of heavy timbering to prevent cave-ins.

The output of the underground mines averages about 275,000 tons monthly.

Total production for the Cerro de Pasco area, amounts to 2'000,000 tons annually, and that from all mines, 4'300,000 tons.

CONCENTRATING

The ores extracted from Cerro de Pasco Corporation mines are generally complex, low-grade sulfides. These ores must be first concentrated and separated into lead, copper and zinc concentrate for further treatment in the smelter. The worthless or gangue material is eliminated in the form of tailings. This process is carried out in our concentrating plants.

In our concentrators, the ores are up-graded by one or more of flotation, gravity and magnetic separation methods. Flotation is based upon certain physical-chemical action of special reagents introduced into a pulp of ground ore. Air injected into this pulp, then forms bubbles, to which the valuable minerals are attached and floated to the surface. This material is collected and is known as a concentrate. The worthless or gangue material is left behind, and is rejected as tailings.

The principle of gravity concentration is based upon separation according to differences in specific gravity, while the magnetic method takes advantage of distinctive magnetic properties of certain materials.

Lead, copper and zinc concentrates are produced by the flotation method, whereas all three methods mentioned above are utilized to produce a tungsten concentrate.

Cerro de Pasco Corporation was one of the first mining companies to successfully carry out lead-zinc flotation and copper-lead separation. The daily tonnage treated in the Corporation's concentrators today may be low in comparison to that of many other mining companies known throughout the world, however none are comparable when it comes to diversification, geographical location of its various mines and concentrators, and many problems created by the complex ores treated.

SMELTIN

Since November 23, 1922, when the bar of copper was cast in La Oroya plant has developed from a smelter produced only blister copper con some gold and silver into a compl lurgical facility which produces lead than any other in South A and more bismuth than any other world.

GOYLLARISQUIZGA

S FLOW-SHEET

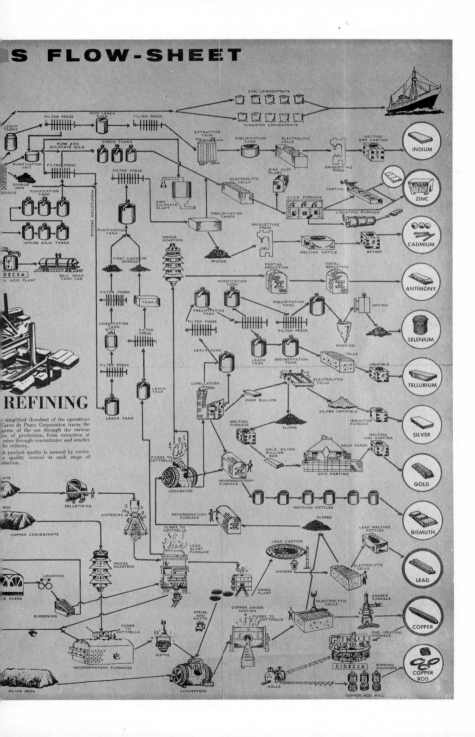

REFING

A simplified flowsheet of the operations
of Cerro de Pasco Corporation traces the
progress of the ore through the various
stages of production, from extraction at
the mine through concentrator and smelter
to the refinery.

High product quality is assured by continuous quality control at each stage of
production.

only the zinc model, which has been extensively tested and used over more than
two years, will be described in detail. The copper and lead systems have been
implemented but are still being tested under routine operating conditions. The
concentrator and mining models have been conceptually developed. A series of
metallurgical plant tests and experiments are being conducted to determine some
principal relationships. The steps that were followed in the model are now
listed.

   i) The determination of the key end products based on a value analysis in
order to capture the largest percentage of revenue with the simplest possible
model.

   ii) The determination of the principal inputs to produce these end products and
identification of all cost elements which vary with the production volume.

iii) The determination of capacity limits or bottlenecks in inputs and outputs.

   iv) The determination of the relationship between inputs and outputs; that is,
how much of each input is required to produce one ton of output.

   v) The determination of accurate variable costs, and coefficients for all
inputs and outputs.

   vi) The determination of other operating policy constraints.

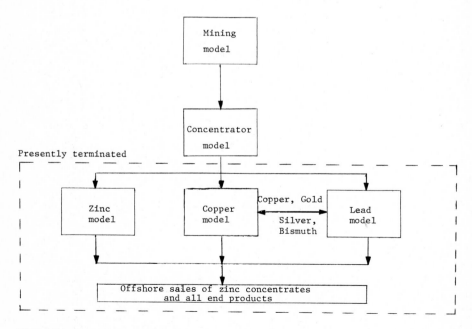

FIGURE 2.   THE OVERALL SYSTEM

Figure 3 shows the flow chart developed in accordance with the first two steps in the copper and lead circuits.

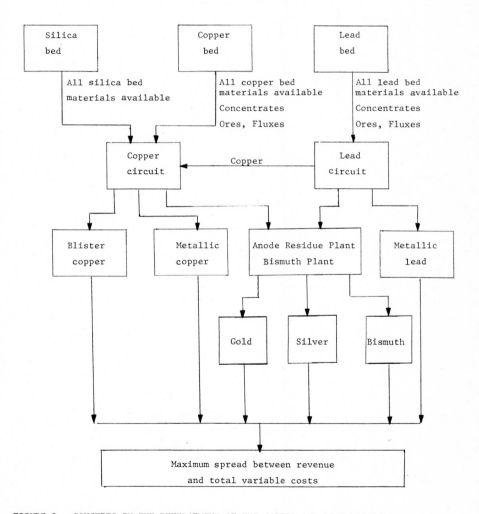

FIGURE 3.   CONCEPTS IN THE DEVELOPMENT OF THE COPPER AND LEAD MODELS

5. *The Zinc Model*

The zinc model covers the upper section of Figure 1. The zinc concentrator produced by the company and purchased from the outside suppliers, are either exported or processed in the La Oroya plant. A total of about 15 types of zinc concentrates are involved, and are exported under contracts with Japanese, American, and European smelters. Each one of these contracts has its own penalty and valuation clauses for impurities and metal assays of each concentrate. The zinc content in a concentrate can vary from 48% to 60%, the silver content can vary from 0 to 16 ounces per ton, cadmium from 0% to 5%, etc. Other impurities, both desirable and undesirable, such as iron, lead, copper, cobalt, etc., also have a wide range in assays. As a consequence of the variety of penalty and valuation of each sales contract, each concentrate can fetch different prices. Shipments have to be made against these contracts on a regular basis, with a maximum and minimum shipment specified by the contract. Each concentrate can also be processed in the company's own smelters in La Oroya.

The concentrates which are used by the company in La Oroya can be fed into two types of roasters. One of these roasters produces a sulfuric gas which is transformed into sulfuric acid, representing one of the end products of the zinc circuit (see Figure 1). The calcine produced by the roasters is chemically processed and filtered in order to separate the soluble zinc sulfate from the leach residue. The zinc sulfate, after purification, is pumped into electrolytic cells which produce the 99.99% pure metallic zinc cathodes, which in turn are cast into slabs of different sizes. The slabs represent the second and most important end product of the circuit. The concentrates processed at La Oroya can be either processed in a Turbulent Layer Roaster (T.L.R.), or in a Fluid Bed Roaster (F.B.R.). Both the roasters would produce calcine. T.L.R., in addition, would produce sulphuric acid. The calcine produced from the roasters would yield zinc, cadmium, and copper to copper beds (see Figure 4).

Any copper obtained in the zinc plant is recycled in the copper circuit. The zinc residue in the zinc plant is processed at another plant to retrieve cadmium and indium. Therefore, there are four key end products in the zinc circuit, namely, zinc slabs, sulfuric acid, cadmium, and indium. The first three products bring 99% of the total revenue in the zinc circuit. Thus, indium is not considered in the model.

A principal decision in the zinc circuit is whether to process or export part or all of the 15 types of zinc concentrates which are input to the circuit. If a part of the concentrates is exported, a decision has to be made concerning the allocation of the total export over various contracts. There is also a limit on the availability of various concentrates and on the magnitude of each sales contract.

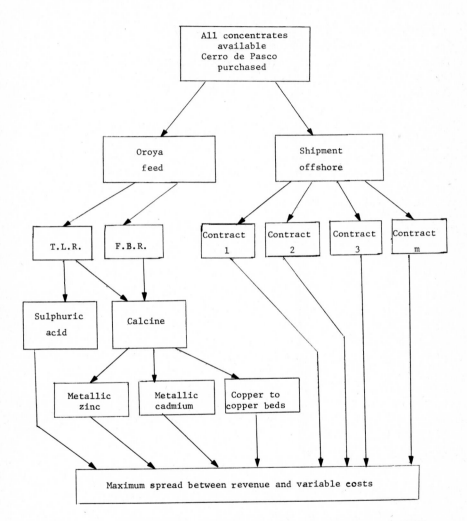

FIGURE 4.   THE INTERRELATIONSHIP BETWEEN DECISION STAGES IN THE ZINC MODEL

    The second set of limitations concerns the zinc plant itself. In particular,
the bottlenecks in the plant at each step in the processing system, were not
known. Therefore, a special study was performed to determine the capacity under
normal operating conditions. These capacities are indicated in Figure 5. For
example, concentrate handling is limited to 25,000 DST (Dry Short Tons) of concentrate per month. All the different processes were expressed in terms of
either the input variables or output variables. For instance, if 12,800 DST of
calcine are leached, they are recomputed in terms of DST of concentrate equivalent to produce this calcine (i.e., 17,294 DST of concentrate equivalent). In
this fashion, all capacities could be expressed on a comparable basis relating
either to input (concentrate tonnage) or output (metallic zinc tonnage). The
smallest tonnage capacity represents the bottleneck and is used in the model as
a maximum limit on inputs or outputs. These are 17,294 DST for the concentrates
(input) at the leaching stage, and 7,312.5 ST (Short Ton) of metallic zinc
(output) in the tank house.

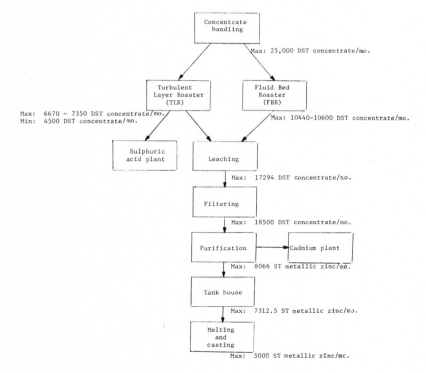

    FIGURE 5. PRODUCTION FLOWS IN THE ZINC CIRCUIT

Through metallurgical tests it had been determined that the total recovery
(percentage of the zinc present in the concentrate which is recuperated as
metallic slab) increases when the zinc assay (i.e., the percentage of zinc present in the concentrate) is higher. A quadratic curve best approximates the behavior of this relationship. The equation is: Recovery = 1.5 $(Zn)^2$, where (Zn)
is the zinc assay of the concentrate fed into the circuit. The cadmium recovered is a simple percentage of the cadmium content of the feed.

Finally, the sulfuric acid production was analyzed as shown in Figure 6.
The Turbulant Layer Roaster (T.L.R.) produces sulfur oxides which are converted
in the sulphuric acid plant into 98.5% sulphuric acid at the rate of .67877 ST
of acid for each DST of concentrate roasted. However, part of this sulphuric
acid is used in the plant itself and, therefore, is not available for sale. As
shown in the diagram the roasters themselves consume .04566 ST of acid per DST
of concentrate. The electrolytic zinc plant uses .07704 ST of acid per ton of
metallic zinc. Cadmium is also a heavy acid user. In addition, the copper and
lead circuits consume a fixed 2,833.4 ST of acid per month. The balance becomes
available for sale.

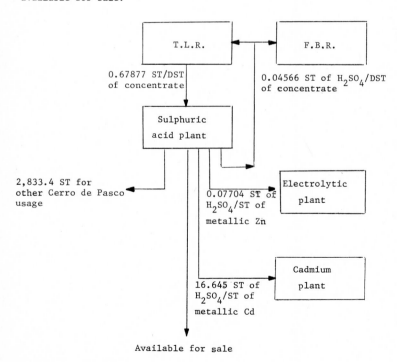

FIGURE 6. SULPHURIC ACID FLOWS

Once all the relationships between inputs and outputs were determined a special cost study was made. The flexible budgeting system which was being implemented at Cerro de Pasco by Cresap, McCormic, and Paget, Inc., supplied the classification of all costs as fixed or variable with changes in production volume. All these costs were recomputed in terms of dollars per concentrate tonnage or end product volume.

Finally, a series of interviews were held with commercial and operating people in order to pinpoint any special constraints. For example, the average iron content in the feed had to remain below 12% in order to keep the processing at high quality standards. Similarly, it was determined that certain off-shore sales contracts would only accept certain types of concentrates.

As a result of these steps the model outlined in Figure 4 is structured in a simplified form in the next section, where only the more significant equations are listed.

6. The Mathematical Formulation

Decision Variables or Unknowns

X_i The number of units of concentrate i processed at La Oroya ($i = 1,2,\ldots,15$).

Y_{ij} The number of units of concentrate i exported under contract j ($i = 1,2,\ldots,15$, $j = 1,2,\ldots,12$).

Z_{i1} The number of units of concentrate i processed through the T.L.R. ($i = 1,2,\ldots,15$).

Z_{i2} The number of units of concentrate i processed through the F.B.R. ($i = 1,2,\ldots,15$).

W_1 The number of units of zinc produced.

W_2 The number of units of cadmium produced.

W_3 The number of units of sulphuric acid available for sale.

W_4 The number of units of copper produced.

Constants or Knowns

a_i The number of units of zinc per unit of concentrate i (i = 1,2,...,15).

b_i The number of units of cadmium per unit of concentrate i
(i = 1,2,...,15).

d_i The number of units of copper recovered from one unit of concentrate i
(i = 1,2,...,15).

e_i The iron content in concentrate i (i = 1,2,...,15).

α The fixed recovery percentage of cadmium.

A_i The number of units of concentrate i available (i = 1,2,...,15).

B_j The number of units of concentrates that can be sold under contract j.

C The number of units of processing capacity available.

D The number of units of the current capacity of zinc production.

E The number of units of the current capacity of cadmium production.

R_1 The revenue from one unit of zinc.

R_2 The revenue from one unit of cadmium.

R_3 The revenue from one unit of sulphuric acid.

r_{ij} The revenue per unit of concentrate i sold under contract j
(i = 1,2,...,15, j = 1,2,...,12).

C The variable cost per unit of concentrate processed.

C_1 The variable cost per unit of zinc produced.

C_2 The variable cost per unit of cadmium produced.

C_3 The variable cost per unit of sulphuric acid sold.

Constraints

From the analysis, notation, and relationships we previously discussed the
following equations are evident.

Zinc production

$$W_1 = 1.5 \sum_{i=1}^{15} a_i^2 X_i \qquad (1)$$

Cadmium production

$$W_2 = \alpha \sum_{i=1}^{15} b_i X_i \qquad (2)$$

Sulphuric acid for sale

$$W_3 = .67877 \ (\sum_{i=1}^{15} Z_{i1}) - .04566 \ (\sum_{i=1}^{15} X_i) - .07704 \ W_1$$
$$- \ 16.645 \ W_2 - 2833.4 \tag{3}$$

Copper to copper beds

$$W_4 = \sum_{i=1}^{15} d_i X_i \tag{4}$$

Availability constraints

$$\sum_{j=1}^{12} Y_{ij} + X_i \leq A_i \quad (i = 1,2,\ldots,15) \tag{5}$$

Sales constraints

$$\sum_{i=1}^{15} Y_{ij} \leq B_j \quad (j = 1,2,\ldots,12) \tag{6}$$

Processing constraints

$$\sum_{i=1}^{15} X_i \leq C \tag{7}$$

Metal production capacity

$$W_1 \leq D \tag{8}$$
$$W_2 \geq E \tag{9}$$

Metallurgical constraints

$$\sum_{i=1}^{15} (e_i - 12) \ X_i \leq 0 \tag{10}$$

Other constraints

$$Z_{i1} + Z_{12} = X_i \quad (i = 1,2,\ldots,15) \tag{11}$$

$$X_i, Y_{ij}, Z_{i1}, Z_{i2}, W_1, W_2, W_3, W_4 \geq 0. \tag{12}$$

Objective Function

The objective function is to maximize the net revenue as described by the following expression.

$$(R_1 - C_1) \ W_1 + (R_2 - C_2) \ W_2 + (R_3 - C_3) \ W_3 + \sum_{i=1}^{15} \sum_{j=1}^{12} r_{ij} Y_{ij} - C \sum_{i=1}^{15} X_i. \tag{13}$$

The linear programming problem is to maximize (13) subject to the constraints (1) through (11) and the nonnegativity requirements (12).

7. *Implementation and Results: The Zinc Model*

The implementation of any study requires the involvement of company personnel. Unless those within the organization accept what the model can do, its implementation would be an impossibility. To assure our success, two of three people in the model team were selected from the corporation.

The model has been run over 200 times in less than a two year period. The runs were made to obtain routine operating plans and also to examine implications of certain decisions. In order to make the zinc model available for repetitive usage, the implementation required development in four directions, namely, automation of the input-output system, installation of a routine feedback information system, model refinements, and the embedding of the model in the company's organizational structure.

The mechanization of the input was accomplished by a computer program which accepts as input, the normal commercial metal quotations, such as the Zinc Producers Price, the Zinc Prime Western Delivered Price for zinc, and the London Bullion Brokers Spot Price for silver. This program updates purchases and sales price lists for all products, and all revenue and cost coefficients.

The output of the linear program is written on a magnetic tape from which an output program extracts relevant information and tabulates them to present to management. The final output contains three sections. The assumption section presents the data used in the model. The operating section gives the best operating plan, and the commercial section prints the best contract-wise allocation of the sale of the concentrates. Sample output is reflected by Figures 7, 8, and 9.

As a necessary requirement for the implementation of the model an information system was developed. As information was received from across the world, certain coordinating centers were established. For example, all price forecasts are made by the New York office as it is most closely informed about the world metal market. Lima, where the purchase of raw materials are negotiated, has to supply data about all purchased materials and services. Finally, the plant operators in the Andes (in La Oroya) have responsibility for the technical data about the metal circuits. A formal feedback system was installed and is shown in Figure 10. With this system, the individual making the forecast also receives the results of the model which are affected by his forecast so that periodic adjustments can be made.

(a)	Zinc	Silver	Cadmium	
PP	20,050		NY	2.6
PWD	18,000	LBDS 162.0	UK	2.6

	Assays				Availability in tons		Cost ($/ton)
	ZN	AG	CD	FE(a)	Maximum	Minimum	
Concentrate A	50.5	3.0	0.16 →		200,000	150,000	115.30
Concentrate B	56.7	9.0	0.19 →		22,000	22,000	131.70
Concentrate C	59.0	3.2	0.10 →		80,000	60,000	120.80
Concentrate D(a)	↓	↓	↓	↓	↓	↓	↓

Notes
(a) Matrix of model continues as needed in both directions.
(b) PP - Producer Price
    PWD - Prime Western Delivered price
    LBDS - London Bullion Brokers price
    NY - New York
    UK - London

Metal prices estimated over 1972 by Mr. J. Doe on 4/23/72

Concentrates supplied for 1972 by Mr. J. Blow on 4/26/72

FIGURE 7.   ZINC MODEL MASTER PLAN FOR 1972:   KEY ASSUMPTION

	Tonnage	Price ($/ton)	Gross revenue($)	Variable cost($)	Contribution($)	Revenue ($/ton) Increase	Decrease
Concentrate intake	153,000			8,800,000		50.50	50.50
Refined zinc produced	63,000			2,200,000		13.60	10.00
Sales							
Metallic zinc							
Local	4,000	330.0	1,320,000				
China	4,000	350.0	1,400,000				
LAFTA	23,000	380.0	8,740,000				
USA	25,000	320.0	8,000,000				
Eurppe	7,000	370.0	2,590,000				
Metallic cadmium	→						
Transportation costs							
Railway (a)				2,100,000			
Total			22,050,000	13,100,000	8,950,000		

Notes
(a) Other cost information is provided in computer printout.

FIGURE 8.   ZINC MODEL MASTER PLAN FOR 1972:   SOLUTION TO PRODUCTION AND MARKETING

	Cerro de Pasco Plant	Contract 1 (USA)	Contract 2 (Germany)	Contract 3 (Japan)	Contract 4 (_____)(a)	Total
Concentrate A						
Price in $/ton	-	118.00	125.00	120.00		-
Optimal tonnage	80,000	0	120,000	0		200,000
Effect on contribution of deviation from optimal						
Increase of 1 ton ($/ton)	(1.20)	12.00	3.50	2.50		3.50
Decrease of 1 ton ($/ton)	(5.80)	-	6.00	1.00		(1.00)
Concentrate B						
Price in $/ton	-	130.00	170.00	139.00		-
Optimal tonnage	0	not acceptable	0	22,000		22,000
Effect on contribution of deviation from optimal						
Increase of 1 ton ($/ton)	(8.20)	10.00	5.00	9.00		5.00
Decrease of 1 ton ($/ton)	-	-	-	3.00		-
Concentrate C						
Price in $/ton	-	125.00	130.00	135.00		-
Optimal tonnage	10,000	10,000	40,000	0		60,000
Effect on contribution of deviation from optimal						
Increase of 1 ton ($/ton)	(1.00)	30.00	5.00	10.00		(5.00)
Decrease of 1 ton ($/ton)	(5.00)	15.00	1.00	8.00		(1.00)
Concentrate D (a)						
Total tonnage maximum	180,000	14,000	170,000	22,000		
Tonnage optimal	180,000	10,000	160,000	22,000		
Revenue/ton increase ($/ton)	7.00	10.00	3.50	(2.00)		
Revenue/ton decrease ($/ton)	15.00	(9.00)	(1.00)	2.00		

(a)Matrix of the model continues as needed in both directions.

FIGURE 9.   ZINC MODEL MASTER PLAN FOR 1972:   SOLUTION TO CONCENTRATE DISTRIBUTION

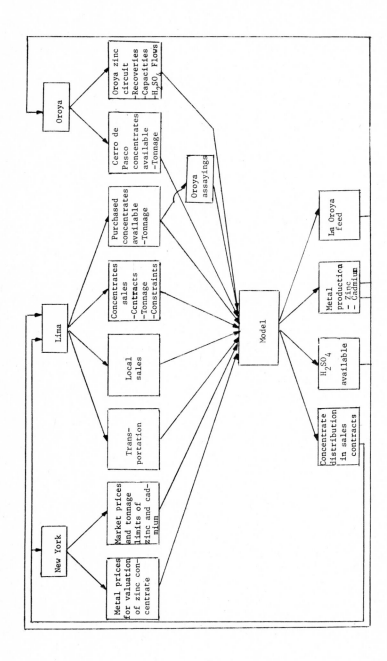

FIGURE 10. INFORMATION FLOW: THE ZINC MODEL

The third type of development in implementing the model relates to more sophistication in the model itself. Other features were added to the model to make it useful for exploring certain non-routine operating problems. For instance, as the electrolytic zinc process consumed very large quantities of electric power, the model was first used to compute the kilowat hours required for each plant. Later on a submodel was added to optimize water usage for power generation in the company's four hydroelectric plants. This is particularly important during seasons when water availability in the Andes is reduced considerably.

In implementing the zinc model a new organization structure for financial reporting was recommended. This would support the decision making process.

The zinc model has been used for preparing routine operating plans. These are used as master guides whenever a significant change occurs in one of the assumptions. Many adjustments to this master plan are, however, made without rerunning the model. For example, when a larger tonnage of concentrate is produced than was expected in the plan, the sales manager can simply consult the concentrate distribution output to decide where it should be allocated. If concentrate B produced 23,000 DST instead of 22,000 DST (Fig. 7) of the plan, Figure 9 shows in the Total Concentrate Column that revenue would increase at $5.0 per DST (Fig. 9). Sales contract 3 is already at its maximum. Sales contract 1 has some room for extra tonnage, but concentrate B is not acceptable under this contract. Contract 2 is the one where the extra DST would be placed. If more than one sales contract is available, which is usually the case, the one with the highest marginal revenue should be chosen. However, the marginal cost data obtained from the LP results are valid over a certain range. In routine reports to management, these ranges are not included. This can be supplied on request.

The concentrate distribution figure also produces useful information for negotiations with the suppliers or customers. If concentrate A is purchased from an outside supplier, the company should not pay more than $1.0 for an additional DST of concentrate A (see Fig. 9).

Similarly, from the marginal cost data in the Total Contract row (Fig. 9), we have that it would be profitable to reduce sales contract 3 and increase the other two contracts. The company can afford paying a penalty of $2/DST to reduce the contract requirements.

Some non-routine strategic problems are solved by using the zinc model. By exploring the consequences of a drop in recovery on the optimal strategy derived with the model, three critical recovery targets were discovered. If the recovery coefficient is below a certain percentage, it is profitable to sell all the concentrates to the customers abroad. If the recovery is moderate, the plant should be used under full capacity. Finally, if the recovery percentage is high, the zinc refinery is used at capacity, but not the roasters. While these results are conceptually logical, the determination of the exact recovery targets which created the shifts of bottlenecks in the system was a significant breakthrough in operating the plant.

The zinc model was also used to evaluate investment proposals. A budgeting of $60,000 for a new grinder in one of the concentrating plants had been turned down on the basis of a criterion used by the corporation's finance department. However, the model helped in determining that this grinder would produce a slightly higher concentrate grade which would allow some substitutions in the zinc plant and in overseas contracts. This would result in over $300,000 additional profit per year.

The model was useful in the evaluation of the maximum price which can be paid for electric power needed to operate the electrolytic cells during the dry season. At these times, hydroelectric power is scarce and by using parametric linear programming all consequences of power shortage is explored.

Finally, contingency plans for strike periods, evaluations of different metal price possibilities, preparations for negotiations with suppliers are now all prepared by using the model.

8. *Conclusions*

The zinc model is only a small segment of the Cerro de Pasco model, which in turn represents only one possible application of corporate short term planning model concepts. However, it helped illustrate the lessons that can be extracted from this experience for other industries.

Three points with some general applicability evolved during this study, and serve as general conclusions.

1. Too many well designed systems fail because they were developed by technical people for technical people. The involvement of the decision makers, the presentation of model results in non-technical terms, the information feedback, and the organizational support are even more important than the purely technical design qualities of a system.

2. The major impact of the model resulted from its scope encompassing the total management viewpoint as opposed to a particular function. A feed mix model, a concentrate allocation model, or a transportation scheduling system, while offering opportunities for improvement, would not have allowed the better understanding of the entire operation.

3. Corporate short term planning models, in industries where some freedom of action exists in terms of raw materials, processes, products, and/or markets presents significant opportunities which are too rarely exploited.

Chapter 3

PETROLEUM PRODUCT DISTRIBUTION PLANNING SYSTEM
BY LINEAR PROGRAMMING

Karl M. Wiig
Arthur D. Little, Inc.
Cambridge, Massachusetts

Abstract. The distribution of petroleum products
from refineries to customers has traditionally
involved a complex evaluation of the best economics,
including complicated exchange arrangements with
competitors. A linear programming model has been
implemented for the distribution system to provide
optimal operating supply routes and exchange levels.
The model handles each customer individually and
also provides operational guidelines on alternate
supply sources for each product to every customer.
A sample problem is presented for the case where
the system includes two refineries, four products,
30 exchange agreements, 150 terminals, and over
1000 customers.

1. *Introduction*

The development of the supply and distribution model that is described in
this chapter was undertaken by a petroleum company to support its marketing,
planning, and operation functions. The model was used to determine the optimal
distribution of product flows given sales forecasts and supply condition options,
and to analyze distribution economics.

The analysis involved the existing distribution system for four major products (premium and regular gasoline, and No. 1 and No. 2 fuel oils) from the refineries to each customer. The costs include the sales prices for each product
and customer, the supply costs, and the product manufacturing costs of the refineries. The economics of sales promotions, competition price allowances, and
other abnormal price changes that were of short duration were not considered in
the analysis. The company desired to include the optimal operation and economic
effects of exchanges of products with other petroleum companies and this became
one of the important features of the planning model.

The problem was formulated as a linear program where a refinery's net return
or ex-refinery's net realization was maximized. It was designed to aid the
analysis of the distribution system by providing:

1. Optimal product distribution paths from refineries and exchangers
    to the customers.
2. Optimal supply terminals for each customer with the next best
    alternative.
3. Relative economic merits of exchange agreements.
4. Guidelines for negotiating exchange agreements.
5. A framework for examining the economic merits of new terminals in the
    system, new customers, and significant exchange transactions.

It was expected that the model would be employed for analysis of the distribution

system on a quarterly or semi-annually basis and that it then should utilize the
latest customer prices, transport costs, and demand forecasts. Hence, the model
was interfaced closely with the present company data files to provide up-to-date
data on an expedient and timely basis.

2. The Problem

The company for which the study was performed is a medium-sized American oil
company with a typical supply and distribution system. It has several refineries
that supply the individual customers (wholesalers, gas stations, etc.) vis common
carrier pipeline systems, private pipelines, and tank trucks through product terminals. A large share of the product is also sold in bulk on the open market to
jobbers, as spot sales or on long-term contracts to other companies. Each refinery
has its own distribution system that is somewhat independent with a small amount of
cross shipments and cross exchanges. The company exchanged about 40% of its products with other oil companies. This is typical for an oil company with a few refineries and with a limited distribution system of their own.

The distribution system supplied over 1000 customers with about 15 categories
of gasolines and fuel oils. The customers could obtain products from more than
150 product terminals, the majority of which were operated by exchange partners and
pipeline companies. The market territory spanned about one half of the continental
United States and products were shipped over four pipeline systems. Products were
exchanged with more than 20 other oil companies utilizing three types of exchange
agreements. The distribution system principally involved the products from one
refinery. However, the company had a purchasing agreement with another oil company. This agreement was represented as an additional refinery. The distribution
system is shown schematically in Figure 1.

Economics

The economics of the distribution system are considered in terms of revenues,
variable costs that are linear with product flows and are in $/gallon. The details
are listed below:

Revenues of sales or product to customers.

Outfreights from product terminals to customers based upon local freight tariffs
and mileage from terminals to customers.

Transportation costs of products from refineries to terminals and between
terminals. These costs are pipeline tariffs (incrementally between terminals
on a pipeline), and direct transportation costs by barge, truck, etc.

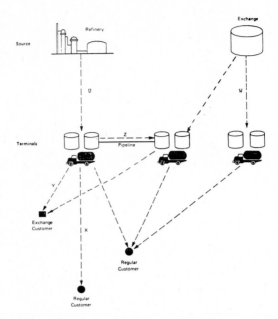

FIGURE 1.   SCHEMATIC REPRESENTATION OF PRODUCT FLOW IN THE DISTRIBUTION SYSTEM.
            The system is a hierarchy in two levels: 1) Product terminals are
            supplied either from refineries, exchange partners, or both;
            2) Customers are supplied from product terminals and so are exchange
            partners when paid back.

Cost of exchange transactions in terms of the marginal cost resulting from the
exchange flows at the different terminal locations.

Product manufacturing costs of each product at the refinery gate. Depending
upon the analysis that is performed, these costs may be marginal product costs
or they may be product transfer prices which then include both variable and
allocated fixed costs.

Sales costs for each product within each sales region.

Exchange Agreements

     The exchange agreements are based upon gallon for gallon exchanges where one
oil company allows customers of a second oil company to draw products from its
terminals. The second company pays back to the first one at another location.
Volumetric imbalances of exchanges between two companies are settled periodically
by return payment in bulk volume in market areas where both companies operate.
Hence, balancing of exchanges involves payment of additional products until the

amounts delivered equals the amounts received. The three exchange types that were
included in this study were:
1. Each exchanged product must be balanced separately at the end of the
    period.
2. Fuel oils and gasolines must be balanced separately.
3. All products are added together and the totals must balance at the
    end of each period.
Exchange agreements formally specify locations where products can be delivered or
received with specification of the economic "place differential" in $/gallon at
those locations. The place differential covers the extra transport costs incurred
by the delivering exchange partner. It is always less than the transport cost that
would be incurred by the receiving exchange partner if he were to supply his own
product to customers at that location from his own refineries (otherwise it will
not pay to have the exchange). In addition to the "place differential", "grade
differentials", and "product differentials" are also specified. Grade differentials specify settlement amounts in $/gallon for differences in product quality
at the two locations. For example, if the product delivered at one location has a
93 octane number, compared to gasoline delivered at the other location with a 92
octane number, the former is a more expensive product and commands a grade differential. Similarly, for exchanges that allow substitution of one product for
another (such as gasoline for fuel oil) a product differential in $/gallon is
specified to allow calculation of the proper settlement amount at the end of a
settlement period. At this time, place, grade, and product differentials are
calculated for all transactions between exchange partners and the resulting economic imbalance is settled by payment.

When there is an unlimited supply, product flows to and from exchange partners are only a function of the amounts that exchange partners have to supply
their customers. In limited supply situations, exchanged volumes become a function
of the terms of the supply agreements, profitability of individual transactions as
measured against all available transactions, and the product availability. The
present analysis principally addresses the unlimited supply situations where local
supply limitations only would occur due to limited transport capacities, etc.

3. *The Model*

The model was formulated as a single time-period linear program to maximize
the net return to the oil company subject to the economic and flow constraints of
the transportation system, the exchange agreements, and the customer demands.

The objective function is given below. The parameters are defined after
the equation.

$$\underbrace{\sum_j \sum_i \sum_r P_{ir} X_{jir}}_{\text{Revenues}} - \underbrace{\sum_j \sum_i \sum_r c_{jir} X_{jir}}_{\text{Outfreight}} - \underbrace{\sum_m \sum_j \sum_r c'_{mjr} U_{mjr}}_{\substack{\text{Direct freight} \\ \text{from refiners}}} - \underbrace{\sum_n \sum_r c''_{rn} Z_{rn}}_{\substack{\text{Pipeline} \\ \text{freight}}}$$

$$+ \underbrace{\sum_t \sum_j \sum_r e_{tjr} W_{tjr}}_{\substack{\text{Exchange} \\ \text{income 1}}} + \underbrace{\sum_j \sum_t \sum_r e'_{tjr} Y_{tjr}}_{\substack{\text{Exchange} \\ \text{income 2}}} - \underbrace{\sum_m \sum_r p'_{mr} V_{mr}}_{\substack{\text{Production} \\ \text{cost}}} - \underbrace{\sum_j \sum_i \sum_r s_{ir} X_{jir}}_{\text{Sales cost}}$$

Known Constants ($/gallon)

P_{ir} The price of product r for customer i.

c_{jir} The outfreight cost of product r from terminal j to customer i.

c'_{mjr} The cost of direct freight of product r from refinery m to terminal j.

c''_{rn} The cost of pipeline freight of product r on pipeline segment n (joining two particular terminals).

e_{tjr} The exchange difference for product r received from exchange t and sent to terminal j.

e'_{tjr} The exchange difference for product r delivered to exchange t from terminal j.

p'_{mr} The cost of manufacturing product r at refinery m.

s_{ir} The cost of selling product r to customer i.

Decision Variables (in 1000 gallons)

X_{jir} The flow of product r to customer i from terminal j.

U_{mjr} The direct flow of product r from refinery m to terminal j.

Z_{rn} The flow of product r through pipeline segment n.

W_{tjr} The flow of product r from exchange t to terminal j.

Y_{tjr} The flow of product r to exchange t from terminal j. (The exchange partner is withdrawing the product.)

V_{mr} The rate of production of product r at refinery m over the time period.

    The constraints in terms of product flows for the system are below. *Customer flow balances* for each customer's (i) demand is required.

$$\sum_j X_{ijr} \leq d_{ir} \qquad \text{for each i and r,} \qquad (2)$$

where d_{ir} is the demand of customer i for product r.

Product terminal flow balances describe all pipeline, truck, and exchange flows into and out of a product terminal. These flows must balance with the inventory change in any time period (see Fig. 2). The model contains one such balance equation for each terminal j and product r.

$$\sum_m U_{mjr} \qquad + \sum_t W_{tjr} \qquad + \sum_{j'} Z_{rj'j} \quad - \sum_{j''} Z_{rjj''} \quad - \sum_t Y_{tjr}$$

(refineries)   (exchanges)   (upstream   (downstream   (exchanges)
                                   terminals)   terminals)

$$- \sum_i X_{jir} \qquad = I_{jr} \qquad \text{for each j and r,} \qquad (3)$$

(customers)

where the symbols are as indicated in Figure 2, and I_{jr} is the change in inventory at terminal j of product r over the period.

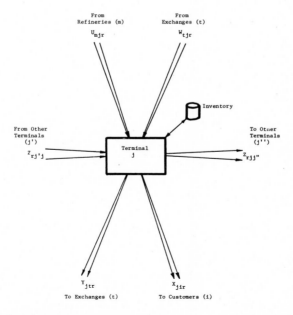

FIGURE 2.   BASIC REPRESENTATION OF PRODUCT FLOWS ASSOCIATED WITH TERMINAL FLOW BALANCE FOR TERMINAL j AND PRODUCT r

Exchange flow balances are required for each exchange t over the time period.
Depending upon the negotiated agreement type, one balance equation will be required
for either each product, each product class (i.e., gasolines and fuels), or only
for the total balance. The latter type of balance is shown below and accounts for
all flows *from* exchange t (W_{tjr}) and all flows *to* exchange t's customers (Y_{jtr}).

$$\sum_j \sum_r W_{tjr} \quad - \sum_j \sum_r Y_{tjr} \quad = b_t \qquad \text{for each } t, \tag{4}$$

(terminals   (terminals
and products)   and products)

where b_t is the change in imbalance for exchange t over the planning period.

Refinery balances for each product r which describes the total flows from any
refinery i to all directly connected terminals is enforced.

$$\sum_j U_{mjr} \quad \leq V_{mr} \qquad \text{for each } m \text{ and } r. \tag{5}$$

(terminals)

Pipeline segment limitations for each segment exist.

$$\sum_r Z_{rn} \leq L_n \qquad \text{for each } n, \tag{6}$$

where L_n is the capacity of pipeline segment n (in 1000 of gallons) over the time
period.

The LP Matrix

The linear programming problem is the objective function (1) subject to the
constraints (2) through (6) and nonnegativity requirements for all the decision
variables. Its coefficient matrix appears as in Figure 3. All balance equations
and customer demands were generated in a staircase fashion as indicated on the
lower right hand corner to allow the use of generalized upper bounding (GUB)
solution procedures. In most test runs, the use of GUB resulted in the suppression
of about 4000 customer demand equations.

The number of explicit constraints and columns in the model can be estimated
using the equation below.

$$\text{Constraints} = N \times R + J \times R + T + T_1 \times R + T_2 \times \frac{R}{2} + T_3 + N^1 + MR$$

Variables	Refinery production variables V_{mr}	Flow to terminals U_{mjr}	Flow from exchanges to terminals W_{tjr}	Flow to exchanges from terminals Y_{tjr}	Pipeline flows Z_{rn}	Customer flows X_{jr}	RHS
Objective function (1)	$-p'_{mr}$	$-c'_{mjr}$	e_{tjr}	e'_{tjr}	$-c''_{rn}$	$(p_{ir} - c_{jir} - s_{ir})$	
Refinery balances (5)	+1 terms	-1 terms					≤ 0
Terminal balances (3)		+1 terms	+1 terms	-1 terms	+1 terms and -1 terms	-1 terms	$= I_{jr}$
Exchange balances (4)			-1 terms	+1 terms			$= b_t$
Pipeline balances (6)					+1 ·terms		$\leq L_n$
Customer balances (GUB rows) (2)						11...1 11...1 ·...1...1	$\leq d_{ir}$

FIGURE 3. THE GENERAL STRUCTURE OF THE COEFFICIENT MATRIX

where N is the number of customers.

      R is the number of products.

      J is the number of product terminals.

      T is the number of exchanges.

      T_1 is the number of exchanges of type 1.

      T_2 is the number of exchanges of type 2.

      T_3 is the number of exchanges of type 3.

      N^1 is the number of pipeline segments that are capacity limited.

and   M is the number of refineries.

A number of the customers do not have demands for all the products, similarly, under normal operating conditions, the majority of pipeline segments are not capacity limited, and, hence, the actual number of constraints estimated by the above equation may be overstated. If GUB is used, the number of GUB rows becomes IR, and this term is omitted from the expression.

    Columns = $N \times R \times N_S + J \times R + T + T \times R \times N_E + T \times R \times N_Y + N \times R + M \times R$,

where N_S is the average number of primary and secondary supply terminals that a customer is assigned.

      N_E is the average number of terminals that are supplied from an exchange partner.

and N_Y is the average number of termianls where an exchange partner may withdraw a product.

4. The Data

A matrix generator program was created to accept data from cards and tape, and to transfer the data into MPS format. The matrix generator program also produced an extensive summary report. The following particular input is accepted by this program.

1. The size of the distribution network.

2. Refinery capacities and product costs.

3. Exchange agreement terms, economics, inventory changes, and exchange partners demands at each location.

4. Pipeline tariffs, capacities, and flow paths.

5. Product terminal supply conditions and inventory changes.

6. Customer demands, sale prices, supply terminal alternatives, and outfreights from each specified terminal to the customer for each product.

Selected samples of the matrix generator summary reports are shown in Figures 4, 5, and 6. Figure 4 shows the refinery data that is used by the model.

Refinery Information ID	Regular	Premium	#1	#2	
No. 1	300000. .1100	100000. .1250	25000. .1000	150000. .0950	Refinery capacity operating costs
No. 2	200000. .1150	120000. .1220	30000. .1010	300000. .0940	Refinery capacity operating costs
No. 3	50000. .1175	25000. .1325	25000. .1100	50000. .1000	Refinery capacity operating costs

Capacity in 1000 gallons over the period.
Product costs in $/gallon.

FIGURE 4. REFINERY DATA SUMMARY (4 PRODUCTS)

Exchange with Midland Oil Company (Agreement No. 76). Basis for Settlement: Type 3 (i.e., Gasolines & Fuels Separately)

Operating Condition = 1 Strict Balance

Regular	Premium	No. 1	No. 2		
				No limit on volumes exchange	
.00	.00	.00	.00	Specified inventory change for each product	(000 gals)
-.00746	-.01766	-.01210	-.00460	Exchange differential *from* exchanger at terminal 765	($/gal)
-.00236	-.01406	-.00850	.00100	Exchange differential *from* exchanger at terminal 896	($/gal)
.00000	.01306	.00750	.00000	Exchange differential *to* exchanger at terminal A48	($/gal)
100.	250.	120.	200.	Demand at terminal A48 (need not be satisfied)	(000 gals)
.00000	.01306	.00750	.00000	Exchange differential *to* exchanger at A69	($/gal)
2300.	600.	150.	1000.	Demand at terminal A69 (need not be satisfied)	(000 gals)

Exchange differentials contain grade, place, and product differentials relative to the basis product.

FIGURE 5. EXCHANGE AGREEMENT SUMMARY

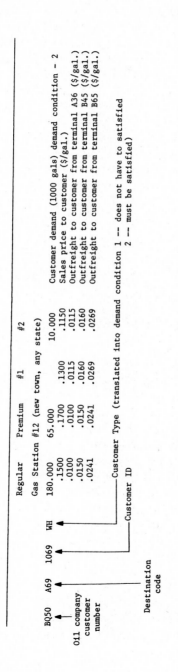

FIGURE 6. REPORT SUMMARY FOR CUSTOMER BQ50

Figure 5 shows consolidated information on one of the exchange agreements. The
place, grade, and product differentials are consolidated, and represented by the
exchange differential for each terminal with a positive figure corresponding to
moneys to be paid by the exchange partner. Figure 6 shows the report for one
customer and indicates the demands, sales prices to the customer, etc.

Data Base

A considerable data base is required for the operation of the planning model.
This data base is maintained regularly, and is tied into the company's general
data processing system. Certain aspects, such as the distribution network topology
and exchange agreement terms are maintained manually, but in punch card form. The
key elements of the data base are:

1. Refinery capacities and manufacturing costs by refinery and product
    (maintained manually).
2. Exchange agreement economics by product terminal, product, and exchange
    partner (maintained manually).
3. Exchange partner demand forecast and historic requirement by product
    terminal and product (maintained as company computer files with optional
    manual update).
4. Pipeline tariffs and capacities (maintained manually).
5. Terminal supply alternatives, economics, and inventories (maintained
    as company computer files).
6. Full customer information (maintained as company computer files).

5. *Implementation, Results*

The model is operated on a Univac 1108 computer utilizing the FMPS linear
programming system which has a generalized upper boudning (GUB) capability. The
general system is shown in Figure 7 which also indicates how the system ties in
with the oil company's general data processing system. Data from the company's
files are used for generation of a "customer tape" with current billing and supply
route and economics information. Another tape with forecasts for customer demands
is also generated externally to the model system and all customer information is
consolidated into a "customer information tape" which is utilized by the matrix
generator and essentially contains information for each customer as was shown in
Figure 6.

The model operation requires considerable computer resource requirements. In
particular, using a Univac 1108, 100k core memory, 700k drum memory, and six tape
drives are required. The model also requires 15 to 60 minutes for each run.

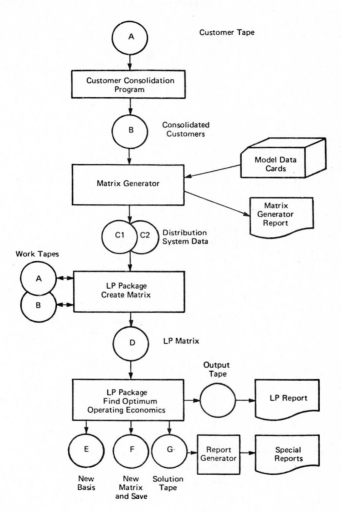

FIGURE 7.

Typical operating statistics and solution times are shown in Tables 1 and 2.

The model is designed to be used primarily as a short range planning tool by periodic execution every three or six months. For this purpose, the model is run with the existing distribution system topology and economics, and with the fore- casts for customer demands for all customers. The model can also be used for analyzing new short and long term business situations such as: new market areas, new exchange opportunities, values of spot sales, and capacity expansion, etc. These situations are analyzed by creating representative scenarios utilizing the data base for the current system as the basis and modifying system topology, economics, and demands.

The model provides extensive information on expected values of operational variables and economics of the system. The detailed LP report is useful for many analysis purposes. Of particular value is the individual customer supply report as shown in Figure 8. This is a sample of LP output with the GUB feature. In this figure, the optimal supply conditions for customer 0108 for regular gasoline (product R) are shown to come from terminal A36, and that this results in a net margin on sale to this customer of $0.007462/gallon and that the next most pro- fitable supply terminal is terminal B51 at a penalty of $0.00437/gallon (which is more than half of the margin). A summary of the general information that is available from the optimal solution is:

1. The net return for sales and distribution of the four products (gasolines and fuels). This value represents sales when no special competitive allowances or promotions are in effect, and where the sales are generated by the customer's forecasted demands.

2. Total revenues for each product under the postulated scenario.

3. Total outfreights from terminals to customers. This amount will be paid to trucking companies or to customers who provide their own trucking.

4. Amounts to be settled with each individual exchange partner at the end of the planning period as a result of place, grade, and product differen- tials. These amounts will be paid to, or are payable by, the exchange partner when balancing at the end of the period. These amounts are functions of the volumes and locations for withdrawn and supplied products.

5. Volumes that are withdrawn from each exchange partner by product and location, and volumes that are supplied to each exchange partner.

6. Optimal and alternate supply origins (product terminal) by product for each customer.

7. Pipeline flows and flows directly from refineries to product terminals by product.

Transportation System Characteristics

    4 Products (R)
    3 Refineries (M)
  150 Terminals (J)
   30 Exchanges (T) ($T_1 = 8$, $T_2 = 15$, $T_3 = 7$)
   50 Pipeline segments (N), 6 with capacity limitations
 1150 Customers (N)
    3 Supply points on the average per customer (N_S)
    4 Terminals supplies per exchange (N_E)
  3.3 Terminals where an exchange withdraws product (N_Y)

Approximate Model Statistics

   725 Rows
  4000 GUB rows (customer demand rows)
 15500 Columns
 39000 Elements
 0.35% Density

Model Estimates by Expressions

GUB rows = 1150 x 4 = 4600
Rows = 150 x 4 + 8 x 4 + 15 x 2 + 7 + 30 + 6 + 3 x 4 = 717 rows
Columns = 1150 x 4 x 3.3 + 150 x 4 + 30 + 30 x 4 x 4 + 30 x 4 x 3 + 60 x 4 +
          + 3 x 4 = 16902 columns

TABLE 1. LP MATRIX STATISTICS (APPROXIMATE)

Matrix generator	8-10 minutes cpu time
LP matrix creation	10 minutes
Revision of matrix	5-7 minutes
Optimum without advance starting point	170 minutes (about 5000 iterations)
Sensitivity analysis from previous basis	
– minor changes	10 minutes
– major changes	60 minutes

TABLE 2. TYPICAL MODEL RUN TIMES (UNIVAC 1108 COMPUTER)

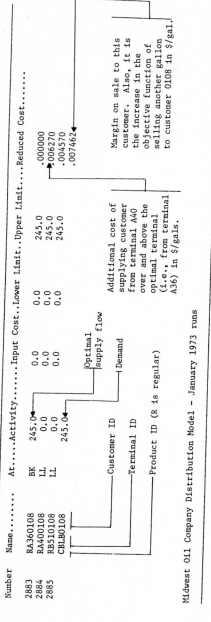

Midwest Oil Company Distribution Model - January 1973 runs

FIGURE 8. CUSTOMER SUPPLY REPORT

6. *Conclusion*

The model has proven quite successful for a distribution system with over 4000
product/customer demands. The model maintenance and periodic solving fall well
within the budget for planning activities that a typical petroleum company of this
size considers reasonable.

The detailed reports from the LP solution on optimal and alternate supply
routes for each customer, with information on the penalties resulting from using
alternate supply routes, provided additional information that otherwise was not
available to the marketing department. The exchange operation report, similarly
consolidated the straight operating economics of the exchanges with the marginal
economics of alternate supply sources and provided exchange operating information
that otherwise had not been obtainable. Hence, the operation of the model, in
addition to providing a framework for testing out proposed system changes such as
introduction of new terminals, provided simple and powerful decision aids for
optimal distribution policies.

References

1. Dantzig, G. B.; Linear Programming and Extensions, Princeton University Press,
    Princeton, 1963, (p. 382).

2. Driebeek, N. J.; Applied Linear Programming, Addison-Wesley Publishing
    Company, Reading, Massachusetts, 1969.

3. Hillier, F. S., and G. J. Lieberman; Introduction to Operations Research,
    Holden-Day, Inc., San Francisco, California, 1967.

4. Hadley, G.; Linear Programming, Addison-Wesley Publishing Company, Reading,
    Massachusetts, 1962 (pp. 273-323).

5. International Business Machines Corporation; Mathematical Programming
    System - Extended (MPSX), Program Description, Manual SH20-0968-1, White
    Plains, New York, 1972.

6. University Computing Company; Functional Mathematical Programming System.
    User's Reference Manual, Dallas, Texas, 1970 (Chapter 2).

Chapter 4

LEASE-BUY PLANNING DECISIONS*

Paul A. Weekes
Xerox Corporation
Stamford, Connecticut   06906

John C. Chambers
Management Sciences Department
Information Services Division
Xerox Corporation
Rochester, New York   14644

and

Satinder K. Mullick
Economics and Operations Research Department
Corporate Plans Division
Corning Glass Works
Corning, New York   14830

---

*This chapter represents a modified version of a paper which originally appeared in *Management Science* 15(6), B295-307 (1969).

Abstract.   A company with a long standing practice
of leasing equipment may amass, over a period of
time, a complicated structure of leasing arrangements,
both long term and short term. Due to changes in
costs of leasing, purchase prices, etc., there should
be frequent evaluations of whether the existing strategy
is optimal. The usual approach is to perform a "macro-
steady-state" analysis, where cash flows and/or return-
on-investment calculations are performed. However, some
macro-analyses indicate only which choice is preferable
and do not take into account either the possibility of
a mixed strategy (own some and lease some) or implementation problems and the resulting costs that occur when
converting from one system to another. Furthermore, if
there is to be a conversion from leased to owned equipment, additional burdens will be placed on the maintenance department.

An overall planning model has been developed which
yields an optimum conversion schedule and the cost of
conversion. The optimal solution can be updated over
time if there are changes in leasing and purchasing
costs, taxes, interest rates, and maintenance costs.
Various assumptions are discussed, along with the inputs,
the model structure, and the use of a matrix generator.
The specific example considered is the problem relating
to leasing or purchasing railroad equipment for transporting finished products to the customer.

1. *Introduction*

In most instances where servicing equipment is used, decisions must be made
as to whether the equipment should be leased or purchased. Almost all companies
have this problem to some extent, whether it be transportation equipment, data
processing equipment, duplicating machines, etc. The usual approach is to perform
a "macro-steady-state" analysis, where cash flows and/or return on investment are
considered to determine whether to own or lease the servicing equipment. However,
some macro-analyses indicate only which choice is preferable and do not take into
account either the possibility of a mixed strategy (own some and lease some) or
implementation problems and the resulting costs that occur when converting from
one system to another. Implementation may take several years to accomplish and a
macro-analysis may show that a change is preferable, while an analysis including
implementation considerations would not indicate that a change is optimal.

Some leasing structures become extremely complicated, evolving over a period
of years, and conversion costs cannot be accurately estimated unless a detailed
optimum conversion schedule is derived. It is therefore important that a model
be developed to determine the optimal conversion schedule, to compute the cost of

conversion and to show what will happen during the conversion period (e.g., fluctuations in maintenance requirements). Finally, the optimal solution may change
over time due to changes in leasing and purchasing costs, taxes, interest rates
(value of money), and maintenance costs.

    While the literature contains a number of papers relating to replacement prob-
lems (Ref. 1), only a few (e.g., Refs. 2, 3, 4, 5, and 6) specifically deal with
lease-buy problems. Furthermore, the writers are not aware of any that discuss
how estimates are obtained relating to the effects of implementation or conversion
on cash flows. This chapter is concerned primarily with a method for deriving
optimum implementation schedules and associated costs, and does not discuss what is
referred to above as the "macro-analysis."

    Some of the basic characteristics of the lease versus buy situation are:

(1)   There is an order lead time for new equipment.

(2)   There is a variety of leasing opportunities and one problem is the
      determination of how long these leases should be.

(3)   There are capacity constraints on maintenance, both for initially
      conditioning the equipment and for regular maintenance.

(4)   The planning horizon is variable and the optimum planning horizon
      should be determined. It is assumed that decisions are reviewed
      over discrete time intervals and the planning horizon is in
      multiples of the review period.

(5)   There are fund limitations at various points in time.

(6)   The purchase price affects the order quantity or vice versa.

Items (5) and (6) have not been considered specifically in the model, but can be
incorporated as additional restrictions.

    The remainder of this chapter describes the development of the model and its
formulation as a linear programming problem, the application of the model to a
specific problem situation, the use of a matrix generator for converting the inputs into the proper form, the presentation of the results of the problem considered, and some of the alternative uses of the model. The objective is *not* to
present the model in detail, but to provide an understanding of how a lease-buy
situation can be handled by the suggested method. While the example given assumes
there is an existing leasing policy, the model is also applicable to the situation
where all items are owned, and new leasing opportunities arise. Also, while the
alternative of renting is not mentioned, the possibility of renting can be treated
by assuming that renting is a short-term lease.

2. *The Model*

The model is verbally described here, while the mathematical development is given in the appendix.

Objective Function

The objective of the model is to minimize, during the planning horizon, for a given service requirement, the sum of the costs of:

(a) owning the servicing equipment,

(b) leasing,

(c) maintenance (including the preparation for placing in service or returning items and having excess equipment during conversion).

Activities (Variables)

The total incremental cost of supplying the required service over the planning period is the sum of the cost of each activity, where the cost of an activity is the product of the number of units and the cost associated with each. The general types of activities that can provide equipment are purchase, new leases, existing leases, and renewing existing leases. The total planning horizon (usually a time span of years) is divided into planning periods (a part of year such as a month or quarter).

Purchasing The opportunity to order purchased items in any planning period comprises the set of all possible purchasing activity. The cost of any particular purchasing opportunity is defined as the equivalent leasing cost per item prorated over the time used (treated as a long term mortgage discounted to present value and considering interest rates, purchase prices, corporate tax rates, annual maintenance costs, usage credit, investment tax credit, salvage value, and depreciation). The estimates of all factors take into account possible future inflation, investment tax credit, changes in leasing rates, etc.

Leasing The model provides the option of entering into new leasing agreements every planning period. The various possible lease lengths from the period equipment must be ordered to the end of the planning horizon constitute the new leasing activities within a planning period. The cost associated with each opportunity is the expected average cost per item for the lease for either a short or long term leasing rate, depending upon the lease length. The existing leases are divided into two categories: (1) normal, and (2) cancellable or option type leases. The model contains a set of activities that represent all existing normal leases and the associated cost is the actual cost per item in a particular lease.

When a normal lease expires, there is a set of activities that represent the renewal opportunities associated with the lease. The structure of these activities
is similar to the new leasing activity, except that the cost is a function of the
particular lease being renewed and the renewal duration (short or long term).

The other category of existing leases is known as cancellable or the option-
type lease. This type is generally characterized by being cancellable on any
yearly anniversary prior to the date of expiration. The number of possible activities associated with each lease depends upon the number of cancellation options
remaining until expiration. The cost of each activity is the actual cost per item
for the particular lease, adjusted for the number of years of the contract not
exercised.

The last set of activities represents the renewal of existing cancellable
leases. The number of renewal opportunities is a function of the number of years
between the lease, expiration and the planning horizon, while the expense related
to each activity is the estimated cost (short or long term) per item for the years
used.

Maintenance and Conversion   Maintenance cost is equal to the average maintenance cost per item owned multiplied by the number of items owned, and the
average maintenance cost per item leased multiplied by the number of items leased.
In addition, there are preparation charges for items newly purchased and those
taken off lease. There is a maintenance capacity constraint, so that some items
may have to be maintained earlier than the time the item is scheduled (required)
to be maintained. The maintenance requirements and capacity constraint mean that
excess equipment, above that required to perform service, must be available if the
service level is to be met. This cost of excess equipment is included in the
ownership and leasing costs.

The objective, as indicated previously, is to minimize the total expected
cost of supplying the required service equipment over the planning horizon. This
is subject, however, to the seven constraints discussed below.

Constraints

(1)   The sum of the items in service in any planning period must be
      equal to or greater than the expected minimum number of items
      required for that period.

          This constraint is represented by an equation for each planning
      period with the variables being the set of all possible equipment
      sourcing activities. The equation coefficient associated with each
      variable, has a value of 1 if items derived from the activity are
      available for service in the respective planning period, otherwise,

the coefficient is zero. The time of availability of items depends
on the number of periods required for delivery (lead time), the time
required to prepare items for service (installation), and the periods
necessary to take leased items out of service and prepare them for
return to the lessor (conversion overlap).

(2) The value of an activity representing an existing normal lease must
be equal to the number of items in that lease.

(3) The sum of the items over the various renewal opportunities associated
with an existing normal lease must not exceed the number of items
covered by the lease.

(4) The sum of the items from the cancellable options associated with a
given lease is equal to the number of items in that existing lease.

(5) The sum of the items over the renewal activities related to a
particular cancellable type lease must not exceed the number of items
that were retained until the expiration of the lease.

(6) The sum of the items that must be processed by the maintenance
facilities in any maintenance period must not exceed the estimated
capacity of the maintenance department for that period. The total
planning horizon plus one year is divided into time increments such
that each segment represents a maintenance planning period. This
constraint has the same form as constraint (1) above in that each
maintenance period is represented by an equation with the variables
being the various equipment sourcing activities. The equation
coefficient has a value of either 1 or 0 depending on whether an
activity places a demand on the maintenance capacity for the
particular maintenance period or not. The demand for maintenance
consists of preparing items to be placed in or removed from service.

(7) The value of every activity must be greater than or equal to zero.

The model can be represented as a linear program with equality and inequality
constraints. As an example, consider the (real) situation below where we wish to
determine the optimum amount of equipment to be purchased and leased at various
points in time throughout the planning horizon.

3. A Case Study

The model was applied to a situation in which a chemical manufacturer main-
tains a large fleet of railroad tank cars for distribution of bulk corrosive
chemicals. Until the study, the company's policy, primarily due to a limitation
of funds for capital investment, was to lease tank cars; consequently, over a
period of years a complicated structure of leases had evolved. Due to the high

demand for rental rail equipment, lease renewal costs had risen sharply and a detailed exploration into the possibility of purchasing equipment rather than leasing was made. An initial "macro-steady-state" analysis indicated that owning
railroad cars was now more economical than continuing to lease them, due to money
being more readily available.

Normally, the implementation program would be to replace all leased cars
whose leases expired with purchased cars until an entire conversion had been
accomplished. However, there were a number of constraints which precluded implementation in this manner. For example, at least nine months of lead time was required to obtain new tank cars. During these nine months some leases would expire
and it was necessary to determine how long the renewal period should be. While
there were financial restrictions, it did not appear that they were a critical
factor in the problem; nevertheless, it was decided that they should be included
in the analysis. The replacement of one tank car required the extensive handling
of two tank cars, since the leased car must be brought to the company's maintenance department for stripping the protective lining and performing other general
maintenance work, and this must be done before the lease expires. Protective
linings must be replaced in any car at intervals ranging from eighteen to twenty
months at a cost of approximately $1,000. Since the linings are perishable and
the leasing companies are not inclined to compete with themselves by selling used
equipment, the alternative of buying leased cars was eliminated from consideration. (The possibility of purchasing the existing leases can be handled by using
this cost as a purchase price of "new" equipment, where there may be several purchase prices with different car lives.) Also, the purchased car must be prepared
for use by the maintenance shop, and hence the replacement rate is limited by the
maintenance shop's capacity. Therefore, leases have to be renewed after a replacement program has been initiated and one part of the implementation plan is to
determine which leases to renew and for what period of time.

The lead time for new equipment includes not only a fixed element due to
preparation and maintenance but also a variable element, since a leased car may
have to be retired from service prior to its expiration date due to the large
turn around time for an individual trip. Through the conversion period there may
exist situations where the capacity to replace equipment exceeds the amount of
leased equipment that expires in any specific maintenance time interval. At times
of excess capacity it may be financially advantageous to acquire additional purchased equipment and bank it to achieve a greater than capacity replacement in
some future period (this strategy was indicated in specific instances in the solution of the actual problem). It can be seen that, because there is a considerable
amount of duplicate equipment during the implementation period, the "optimum" solution, considering only conversion cost, may no longer be the best solution.

Another further complication is the nature of the present and future lease.
There were two types of lease contracts: (1) a normal contract that has a fixed
termination date, and (2) a "cancellable" lease that can be terminated on any
yearly anniversary prior to the lease's date of expiration. The expiration dates
range over several years and individual contracts vary greatly both in the leasing
cost per car and the number of cars. There are two basic types of new or renewed
leases, long and short term. A long term contract is for three or more years and
generally carries lower lease rates, whereas a short term contract is for one to
two years with lease rates being higher. Also, a normal increase in sales level
requires an increase in the number of cars required and maintenance requirements
over time. Finally, both financial conditions and purchase costs change, as well
as other factors, so that the feasibility of entering into new leasing agreements
must be frequently examined. The example below describes how the general model
was applied to the above.

Example

The actual situation includes over 100 existing leases, involving approximately 1200 railroad tank cars. The planning horizon was five years, divided into
quarters, so that the resulting linear programming model exceeded 200 rows and 500
columns, not including slack variables. For illustrative purposes, a small sample
problem has been formulated to demonstrate the construction of the tableau. The
size of the problem has been reduced, but all of the restrictions are included.

Table 1 contains the time parameters that must be specified to formulate the
linear programming model. The planning horizon has been limited here to two years
(8 periods) and leases of two or more years are considered long term.

Time Data

Time period (i)	.25 years (1 quarter)
First model period	1-68 (quarter year)
Planning horizon (I)	8 periods (2 years)
Purchase lead time	2.5 periods
Item preparation	.5 periods
New lease lead time	2.5 periods
Return preparation	.5 periods
Maintenance capacity period (t)	.5 years
Long term lease	2 years or more

TABLE 1. TIME PARAMETERS

The purchase and new lease cost data for the sample problem are presented in Table 2. For simplicity, the purchase cost data is not given in raw form (i.e., the way it would be collected), but as the equivalent lease value, calculated using equation (2) in the appendix. All costs are in terms of dollars per item period.

Period	1	2	3	4	5
Equivalent lease value (purchase cost)	$275	$275	$275	$300	$300
Short term	$300	$325	$325	$350	$375
Long term (new lease cost)	$250	$275	$300	$325	$325

TABLE 2.   PURCHASE AND NEW LEASE COST DATA

Four typical existing leases are included in the sample problem, two normal and two of the cancellable or option type. The descriptive data concerning the present leases are shown in Table 3.

Type	Lease #	Items	Expiration Date	Present Cost	Expected short term renewal cost	Expected long term renewal cost
Z	1	10	1–69	$300	$350	$325
Z	2	15	3–68	$275	$340	$315
V	1	12	4–69	$310	$375	$350
V	2	17	1–70	$300	$360	$340

TABLE 3.   PRESENT LEASE INPUTS

The final set of input data required, which are the constraints that are given in Table 4, is the minimum number of items required to be in service in each planning time period and the maximum number of items that can be handled by the maintenance shops during each maintenance time period.

Period	1	2	3	4	5	6	7	8
Minimum number of items	54	54	54	58	60	65	65	70
Maintenance capacity	36	36	40	45	50	50	--	--

TABLE 4.   MODEL CONSTRAINTS

The data given in Tables 1 through 4, must first be converted to a form that can be used as input for the linear programming model. Considerable computations are required and a *matrix generator* is used to convert the input data into the proper form. The general logic of the matrix generator program is illustrated in Figure 1. This program generates, vector by vector, every possible activity

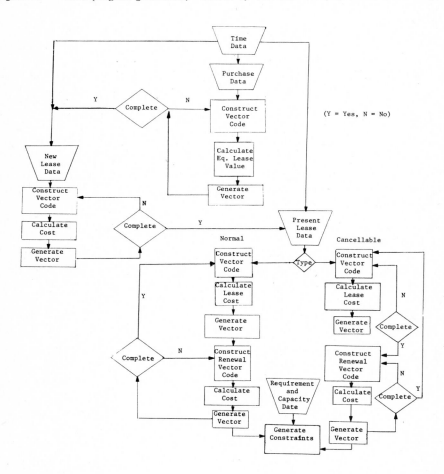

FIGURE 1. GENERAL LOGIC OF THE MATRIX GENERATOR

that can take place within the specified planning horizon. (Vectors are constructed for each activity as defined in the appendix.) All matrix entries are generated and cost calculations are performed as specified in the model. This

special purpose matrix generator uses a somewhat complicated coding structure to
identify each row and vector. In the case of vectors, in addition to identifying
the type, numeric codes are used to indicate such parameters as lease duration,
period of order, period of delivery, first service period, termination period,
preparation maintenance period, and return maintenance period. The coding scheme
is useful in sorting the solution output to form various schedule reports. The
matrix generator concept has several advantages; namely, the elimination of much
manual coding, the avoidance of clerical and calculation errors, and the ease of
updating the model by changing input data.

To illustrate the function of the matrix generator in constructing the linear
model, a representative vector subset is presented in Figure 2. The two vectors

Variable	Y_{11}	Y_{12}	
Cost	$1200	$1430	
Period 1			
Period 2			
Period 3			1 means
Period 4	1	1	equipment
Period 5	1	1	in service
Period 6	1	1	
Period 7		1	
Period 8		1	
Maintenance Period 2	1	1	
Maintenance Period 3			1 means
Maintenance Period 4	1		equipment requires
Maintenance Period 5			maintenance
Maintenance Period 6		1	

FIGURE 2. VECTOR STRUCTURE FOR THE SAMPLE PROBLEM

(Y_{11}, Y_{12}) represent the activities of entering into new leasing agreements,
where the first subscript is the planning period in which the equipment is
ordered, and the second subscript indicates the lease duration in years. Since
the duration of Y_{12} in the example exceeds the end of the planning horizon, no
further new lease vectors whose first subscript is 1 will be generated.

Although the equipment is ordered in Period 1, it does not appear in service
until Period 4 (indicated by the unit matrix entries in Figure 3). The three-
period delay is the sum of the 2.5 period delivery lead time and the .5 period
item preparation time (Table 1).

variables	X1	X2	X3	X4	X5	Y11	Y12	Y21	Y22	Y31	Y41	Y51	Z1	Z2	U11	U21	U22	V10	V11	V20	V21	W11	
costs	1516	1238	963	750	450	1200	1430	1300	1260	1300	870	500	1050	686	1400	1360	2160	2325	1095	2720	1440	500	
Period 1													1	1				1	1	1	1		≥ 54
Period 2													1	1				1	1	1	1		≥ 54
Period 3													1			1	1	1	1	1	1		≥ 54
Period 4	1					1	1						1			1	1	1		1	1		≥ 58
Period 5	1	1				1	1	1	1						1			1	1	1			≥ 60
Period 6	1	1	1			1	1	1	1	1					1			1	1	1			≥ 65
Period 7	1	1	1	1			1	1	1	1	1				1			1		1			≥ 65
Period 8	1	1	1	1	1		1		1	1	1	1			1			1		1		1	≥ 70
Lease Z1 Cont.													1										= 10
Lease Z2 Cont.														1									= 15
Lease V1 Cont.																		1	1				= 12
Lease V2 Cont.																				1	1		= 17
Renewal Cont. Z1															1								≤ 10
Renewal Cont. Z2																1	1						≤ 15
Renewal Cont. V1																		-1				1	≤ 0
Maint. Period 2	1	1				1	1	1	1				1			-1	-1		1		1		≤ 36
Maint. Period 3			1	1						1	1		1		-1								≤ 40
Maint. Period 4				1	1						1					1			1			-1	≤ 45
Maint. Period 5											1	1				1		1					≤ 50
Maint. Period 6							1			1		1						1				1	≤ 50

FIGURE 3.   SAMPLE PROBLEM

In Figure 3 and in the remainder of this chapter the following notation is used.

> X represents the purchase activities.
>
> Y represents the new leasing activities.
>
> Z represents the existing normal leases.
>
> U represents the renewal of normal leases.
>
> V represents the existing cancellable leases.
>
> W represents the renewal of cancellable leases.

Since items obtained from lease Y_{11} are delivered in the middle of Period 3, (Fig. 2), they must be returned in the middle of Period 7 since the lease duration is one year. Also, the first half of Period 7 is required for the preparation and return of the rental equipment. (Reconditioning must be performed before the end of the lease term.)

The cost associated with Y_{11} ($1200) is the short term cost per period ($300) times the lease duration in periods (4 periods = 1 year). Since Y_{12} has a duration that exceeds that of the planning horizon, the associated cost for that new lease opportunity is defined as the total lease cost (long term) per item ($250, Table 2), divided by the number of possible service periods (7, Fig. 2) times the model service periods (5, Fig. 2) or:

$$\frac{(8 \times \$250)}{7} \times 5 = \$1430.$$

The unit entries in the maintenance capacity rows (Fig. 2) indicate the maintenance periods in which either preparation or return maintenance must be performed. It should be noted that the maintenance capacity planning period exceeds the
service planning horizon to prevent the occurrence of unrealistic solutions.

The other types of activities are similarly structured and the entire sample
tableau as obtained from the matrix generator is presented in Figure 3. The output of the matrix generator is therefore in the proper form to be accepted by a
standard linear programming package and the solution, in terms of vector activity
levels, is given in Table 5.

Vector	X_2	X_3	X_4	X_5	Y_{12}	Z_1	Z_2	U_{21}	V_{10}	V_{11}	V_{20}
Value (no. of items)	12	5	15	7	14	10	15	15	2	10	17

TABLE 5. SAMPLE PROBLEM SOLUTION

The solution is presented in a more usable form in Table 6. The equipment
derived from new leases and purchasing activities, the disposition of existing
leases, and the distribution of leased and owned equipment over the planning
period are shown.

4. *Alternative Uses of the Model*

It has been found that the model, while developed primarily for a detailed
analysis of lease-buy situations, has a variety of alternative uses. First, it
is quite likely that the strategy of converting from leasing to owning might force
the leasing company to re-examine its leasing rates. The optimal solution assuming that the leasing company does not lower rates will be different as compared to
the optimal solution assuming that the leasing company will lower its rates.
Hence, the model should be used to re-evaluate strategies as the alternatives or
parameter values change. A similar type of analyses are the sensitivity analyses
to determine under what conditions the optimal solution changes. For example,
the optimal solution may be sensitive to investment credit, mileage credit, and
interest rates. Furthermore, these factors should be tracked regularly and, if
necessary, the optimal strategy with respect to the present state should be
modified or changed.

The model solution outputs may serve as inputs to a game theory type model,
where reactions of the leasing companies to proposed purchases are taken into
consideration in deriving the optimal solution. Probabilities of leasing rate

Period	1	2	3	4	5	6	7	8
Purchase equipment								
ordered		12	5	15	7			
in service					12	17	32	39
New lease equipment								
ordered	14							
in service				14	14	14	14	14
Lease 1 equipment (Z_1)								
returned					10			
in service	10	10	10	10				
Lease 2 equipment (Z_2)								
returned							15	
renewed			15					
in service	15	15	15	15	15	15		
Lease 3 equipment (V_1)								
cancelled				10				
returned				10				2
in service	12	12	12	2	2	2	2	
Lease 4 equipment (V_2)								
in service	17	17	17	17	17	17	17	17
Equipment in service								
leased	54	54	54	58	48	48	33	31
owned					12	17	32	39
total	54	54	54	58	60	65	65	70

Total cost $160,021.00

TABLE 6.   SAMPLE PROBLEM SOLUTION (no. of items)

changes and changes in other factors can be included directly in the model, or other game theory approaches used.   In the case of railroad tank cars, the question of obsolescence was not critical.   However, in the case of computers or similar equipment, where the technological innovations are taking place rapidly, the probability of obsolescence over time must be considered in determining an optimum policy.

5. *Conclusion*

When lease-buy situations are evaluated, it is frequently important to consider, in detail, the implementation costs and related problems.   Also, a detailed

implementation model may be used not only for decision making based on initial
data, but when new alternatives arise or conditions change. The detailed model
can be constructed so that linear programming methods can be applied to derive
solutions. Finally, it has been shown that the use of a matrix generator will
considerably reduce the processing requirements for input data.

APPENDIX: *The Mathematical Description of the Model*

This section is structured (definition of costs, etc.) as the verbal description in Section 2 (The Model).

The cost equation for the total incremental cost of supplying the required
service over the planned period, expressed as the sums of the products of a number
of vectors may be stated as

$$CX + C^+Y + C''Z + DU + D^+V + D''W, \tag{1}$$

where each of the terms are defined below.

C is the expected cost per item purchased (known).
X is the number of items purchased (unknown).
C^+ is the expected cost per new item leased (known).
Y is the number of new items leased (unknown).
C'' is the actual cost of an existing lease (known).
Z is the number of existing normal items leased (unknown).
D is the expected cost per renewed item leased (known).
U is the number of renewed normal items leased (unknown).
D^+ is the actual refundable cost per existing item leased (known).
V is the number of existing leased items that are cancellable (unknown).
D'' is the estimated cost of renewing an existing item leased (known).
W is the number of existing cancellable items leased that are renewable (unknown).

As indicated previously, the existing leases are broken into two types; the
set of all existing normal leases (Z) and the set of all existing cancellable or
option type leases (V). The DU and $D''W$ terms in equation (1) represent the total
costs incurred from the renewal of each type lease, where DU is the total cost of
a normal lease renewal activity, and $D''W$ is the total cost of a cancellable lease
renewal activity.

The vector notation for each of the terms are described below.

$$X = \begin{pmatrix} X_1 \\ X_2 \\ \vdots \\ X_h \\ \vdots \\ X_H \end{pmatrix}$$

The set of all possible purchase activities, where X_h is the number of items ordered in the hth period. (The total planning horizon is divided into I increments of time such that $i = 1, 2, \ldots, I$.) In general, $h = i$ but $H < I$ since items ordered later than the Hth period could not be delivered and prepared for service by the end of the model planning period. (Unknown)

$C = (c_1, c_2, \ldots, c_h, \ldots, c_H)$, where c_h is the expected leasing cost per item for the (X_h) purchase activity. (Known)

To calculate the expected leasing cost (equation (1)) of a purchased item, the first step is to convert the estimated purchase price $(\hat{P})$ to an annual (year n) payment $(f(r,N)\hat{P})_n$ assuming an (N) year mortgage at an estimated interest rate (r). The annual (year n) cash flow is computed by adding the estimated maintenance cost $(\hat{M}_n(t))$, which is borrowed by the lessor under a leasing agreement, to the mortgage payment and subtracting the estimated purchased item usage credit $(\hat{R}_n)$, investment tax credit $(\hat{C}_o)$ in the first year, and the estimated residual value at the end of the mortgage $(\hat{G}_N)$. This sum is decreased by tax deductions on the annual (year n) interest paid $(\hat{I}_n)$, the maintenance costs, and the annual (year n) depreciation $(\hat{D}_n)$, and increased by the tax liabilities on usage credit received and recaptured residual value at the end of the mortgage at an estimated tax rate $(\hat{T})$. Each annual cash flow is discounted to present value $((1+\hat{A})^{-n})$ using an estimated discount rate $(\hat{A})$, and the total discounted cash flow is calculated by summing these over the life of the mortgage $(\Sigma_{n=0}^{N})$. The total discounted cash flow is then converted back to an annual payment using the present value discount rate or the corporate value of money $(f(\hat{A}, N))$. To make the estimated annual payment equivalent to a leasing cost, the estimated annual usage credit for a leased item (R) is added and the sum is multiplied by $(1/(1-\hat{T}))$ since the annual leasing cost is tax deductible. Finally, the annual expected leasing cost is multiplied by the number of years (L_h) from delivery to the end of the planning horizon to arrive at the unit cost (c_h) for a particular purchase activity. All values are estimated with respect to the time an order is placed. This explains equation (2).

$$c_h = \left(\frac{f(\hat{A}, N)}{1-T} \left(\sum_{n=0}^{N} (1+\hat{A})^{-n} \{ [f(r,N)\hat{P}]_n + \hat{M}_n(t) - \hat{R}_n - \hat{C}_o - \hat{G}_N \right.\right.$$
$$\left.\left. - \hat{T}[\hat{I}_n + \hat{M}_n(t) + \hat{D}_n - \hat{R}_n - \hat{G}_N] \} \right) + \hat{R} \right) L_h, \qquad (2)$$

where all terms are known input values and summarized on the next page.

$\hat{A}$ is the estimated present value discount rate.

N is the number of years for which equipment will be mortgaged.

$f(\hat{A},N)$ is the annual payment coefficient including simple interest at a rate $(\hat{A})$
for (N) years; i.e., $f(\hat{A},N) = \hat{A}(1+\hat{A})^N/[(1+\hat{A})^N - 1]$.

$\hat{T}$ is the estimated corporate tax rate.

$f(r,N)$ is the annual payment coefficient including simple interest at a rate (r)
for (N) years; i.e., $f(r,N) = r(1+r)^N/[(1+r)^N - 1]$ (e.g., $f(.055,20) = .08368$).

r is the estimated mortgage interest rate.

$\hat{P}$ is the estimated purchase prices per item.

$\hat{M}_n(t)$ is the estimated maintenance cost per item in year n. $\hat{M}_n$ is a function of
time or item age.

$\hat{R}_n$ is the estimated average usage credit received per item purchased in year n.

$\hat{C}_o$ is the estimated investment credit per item in the 0th year.

$\hat{G}_N$ is the estimated residual value per item at the end of the mortgage.

$\hat{I}_n$ is the interest payment per item in year n.

$\hat{D}_n$ is the estimated depreciation taken per item in year n.

$\hat{R}$ is the estimated average usage credit per year for an item leased.

L_h is the number of years from delivery to the end of the model for the
purchase activity corresponding to X_h (the number of items involved in the
h^{th} period).

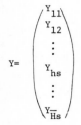

$Y=$
$$Y = \begin{pmatrix} Y_{11} \\ Y_{12} \\ \vdots \\ Y_{hs} \\ \vdots \\ Y_{\overline{Hs}} \end{pmatrix}$$
The set of all new leasing activities. Whereas y_{hs} is the number
of items ordered in the hth period having a lease duration of s
years. H is a function of the delivery lead time for new leased
equipment and is a function of h or the remaining model years.
(Unknown)

$c^+ = (c^+_{11}, c^+_{12}, \ldots, c^+_{hs}, \ldots, c^+_{\overline{Hs}})$, where c^+_{hs} is the average expected cost per item
for the model life of the Y_{hs} activity, either short or long term, depending on s.
(Known)

$$Z = \begin{pmatrix} z_1 \\ z_2 \\ \vdots \\ z_q \\ \vdots \\ z_Q \end{pmatrix}$$
The set of existing normal leases, where q is an arbitrary lease identification number. (Unknown)

$C'' = (c_1'', c_2'', \ldots, c_q'', \ldots, c_Q'')$, where c_q'' is the actual cost per item of the Z_q lease.

$$U = \begin{pmatrix} U_{11} \\ U_{12} \\ \vdots \\ U_{qs} \\ \vdots \\ U_{Qs} \end{pmatrix}$$
The set of normal lease renewal activities, where U_{qs} is the number of items from the Z_q lease that are renewed for s years. The maximum value of s for any q depends on the expiration period of the Z_q lease. (Unknown)

$D = (d_{11}, d_{12}, \ldots, d_{qs}, \ldots, d_{Qs})$, where d_{qs} is the average expected cost per item for the model life of the U_{qs} activity. The cost is a function of the lease being renewed and the renewal duration (short or long term). (Known)

$$V = \begin{pmatrix} V_{10} \\ V_{11} \\ \vdots \\ V_{jk} \\ \vdots \\ V_{Jk} \end{pmatrix}$$
The set of existing leases that can be cancelled on any yearly anniversary prior to the date of expiration, where V_{jk} is the number of items in the jth existing cancellable lease that are cancelled k years before the expiration date of the lease. (Unknown)

$D^+ = (d_{10}^+, d_{11}^+, \ldots, d_{jk}^+, \ldots, d_{Jk}^+)$, where d_{jk}^+ is the actual cost per item of the jth lease adjusted for the k years of contract not used. (Known)

$$W = \begin{pmatrix} W_{11} \\ W_{12} \\ \vdots \\ W_{js} \\ \vdots \\ W_{Js} \end{pmatrix}$$

The set of all possible activities representing the renewal of the cancellable leases (V), where W_{js} is the number of items from the jth lease that are renewed for s years. (Unknown)

$D'' = (d''_{11}, d''_{12}, \ldots, d''_{js}, \ldots, d''_{Js})$, where d''_{js} is the average estimated cost per item of renewing (short or long term) the jth lease for s years, over the model life of the W_{js} activity. (Known)

The objective, as indicated previously, is to minimize the total expected cost equation (1), subject, however, to the following seven constraints.

(i) The sum of the items in service in any planning period must be greater than or equal to the expected minimum required number of items for that period. Or, using our notation,

$$AX + A^+Y + A''Z + BU + B^+V + B''W \geq F. \tag{3}$$

Each term on the left hand side is the product of an activity *vector* and a coefficient *matrix*. Any element of the coefficient matrix has a value of 1 if items from the associated activity are in service in the indicated period, otherwise the element has a value of 0. In particular (all terms are known),

$A = (a_{ih})$ where $a_{ij} = 1$ if $i \geq h$ plus the number of periods required for delivery and item preparation of the X_h activity. (A has I rows and H columns.)

$A^+ = (a^+_{ihs})$ where $a^+_{ihs} = 1$ if $i \geq h$ plus the number of periods required for delivery and equipment preparation and also does not exceed the periods that the (Y_{hs}) lease expires minus the periods required to take the leased items out of service and prepare them for return to the lessor. (A^+ has I rows and $\overline{H}_s$ columns.)

$A'' = (a''_{iq})$ where $a''_{iq} = 1$ for i less than or equal to the expiration period of the (Z_q) lease minus the number of periods required to take the leased items out of service and prepare them for return. (A'' has I rows and Q columns.)

$B = (b_{iqs})$ where $b_{iqs} = 1$ for i greater than the expiration period of the Z_q lease minus the periods of the lease that would be lost if the items were not renewed and also i does not exceed the expiration period of the (U_{qs}) renewal contract minus the periods needed to prepare the items for return. (B has I rows and Qs columns.)

$B^+ = (b_{jk}^+)$ where $b_{jk}^+ = 1$ for i less than or equal to the cancellation period of the (V_{jk}) activity minus the time required to prepare the items for return. (B^+ has I rows and Jk columns.)

$B'' = (b_{js}'')$ where $b_{js}'' = 1$ for i larger than the expiration period of the V_{jo} activity minus the periods that would be lost if the lease was not renewed and also i does not exceed the expiration period of the (W_{jo}) renewal activity minus the periods needed to prepare the items for return. (B'' has I rows and Js columns.)

$$F = \begin{pmatrix} f_1 \\ f_2 \\ \vdots \\ f_i \\ \vdots \\ f_I \end{pmatrix}$$

where f_i is the estimated minimum number of items for the ith period.

(ii) The value of existing leased items in a given lease must be equal to the number of items in that lease. Or, we have

$$Z = E, \tag{4}$$

where (E is known)

$$E = \begin{pmatrix} e_1 \\ \vdots \\ e_q \\ \vdots \\ e_Q \end{pmatrix}$$

This requires that the value of the Z_q activity be equal to the number of items in the qth lease (e_q). (Known)

(iii) The sum of the items from a given lease (Z_q) that are renewed does not exceed the number of items in that existing lease. Thus,

$$\sum_s U_{qs} \leq e_q \qquad \text{for each q.} \tag{5}$$

(iv) The sum of the items from the cancellable options associated with a given lease is equal to the number of items in that existing lease. Therefore,

$$\sum_k V_{jk} = e_j \qquad\qquad \text{for each } j. \qquad\qquad (6)$$

(v) The sum of the items from a given lease that are renewed does not exceed the number of items that were kept until the contract expiration for that lease. In our notation,

$$\sum_s U_{js} = V_{jo} \qquad\qquad \text{for each } j. \qquad\qquad (7)$$

(vi) The sum of the items that must be handled in any maintenance period must not exceed the estimated capacity of the maintenance shops for that period. Using vector notation, we have

$$PX + P^+Y + P''Z + GU + G^+V + G''W \leq M. \qquad\qquad (8)$$

The total planning horizon, plus one year, is divided into T time increments such that $t = 1,2,\ldots,T$, where t is a maintenance period. Equation (8) has the same form as equation (3), where the elements of the coefficient matrices assume a value of +1 if items from the various activities must be handled in period t. In particular (all terms are known),

$P = (p_{th})$ where $p_{th} = 1$ if t is equal to the period in which the items from the X_h activity must be prepared for service. (P has T rows and H columns.)

$P^+ = (p^+_{ths})$ where $p^+_{ths} = 1$ if t is equal to the period when items from the Y_{hs} activity must be prepared for service and when t is equal to the period when items from that lease must be prepared for return. (P^+ has T rows and $\overline{H}s$ columns.)

$P'' = (p''_{tq})$ where $p''_{tq} = 1$ if t is the period that the Z_q lease must be prepared for return. (P'' has T rows and Q columns.)

$G = (g_{tqs})$ where $g_{tqs} = 1$ if t is the period that the items from the U_{qs} activity must be prepared for return; $g_{tqs} = -1$ when t is equal to period that items from the Z_q lease would have been prepared for return if the lease was not renewed. (G has T rows and Qs columns.)

$G^+ = (g^+_{tjk})$ where $g^+_{tjk} = 1$ if t is the period when the items from the V_{jk} activity would have to be prepared for return. (G^+ has T rows and Jk columns.)

$G'' = (g''_{tjs})$ where $g''_{tjs} = 1$ if t is the period that items from the W_{js} activity must be prepared for return; $g''_{tjs} = -1$ if t is the period when the items from the V_{js} activity must be prepared for return. (G'' has T rows and Js columns.)

$$M = \begin{pmatrix} m_1 \\ m_2 \\ \vdots \\ m_t \\ \vdots \\ m_T \end{pmatrix}$$
where m_t is the estimated maintenance capacity for the tth maintenance period.

(vii) The value of every activity must be greater than or equal to zero; or

$$U,V,W,X,Y,Z \geq 0. \tag{9}$$

The linear program, consisting of the objective function (1), constraints (3) through (8), and nonnegativity requirements (9), may be posed in the format suitable for solution by any linear programming system. Note that all the variables (i.e., U, V, W, X, Y, and Z) should really be integer constrained, however, in all applications to date an integer solution was obtained. Of course, this phenomenon may not occur, and heuristics and/or integer programming techniques may have to be used.

References

1. Ackoff, R. L.; Progress in Operations Research, Vol. I, John Wiley and Sons, New York, 1961.

2. Hata, K.; "A Decision Curve for Lease or Buy," Management Services, January - February, 1967.

3. Kirby, M. W.; "Determining the Optimum Size of Rented and Owned Fleets," paper presented at the annual meeting of the Operations Research Society, Western Section, Honolulu, Hawaii, September 1964.

4. Shephard, R. W.; "Lease Cost Scheduling of Daily Local Deliveries; An Application of Dynamic Programming," Production Engineering 39(11), 629-637 (1960).

5. Taylor, G. A.; Managerial and Engineering Economy, D. Van Nostrand Co., Inc., Princeton, N. J., 1964.

6. Whisler, W. D., "A Stochastic Inventory Model for Rented Equipment," Management Science 13(9), 640-647 (1967).

Chapter 5

OPTIMAL PRODUCTION SCHEDULING AND INVENTORY CONTROL
UNDER NON-STATIONARY DEMANDS

N. K. Kwak*
School of Business and Administration
Saint Louis University
Saint Louis, Missouri 63108

and

Stephen A. DeLurgio
School of Business Administration
Sangamon State University
Springfield, Illinois 62708

*We are indebted to our colleague, Charles B. Drebes, for the helpful comments.

Abstract. This paper examines and analyzes the optimal planning and scheduling of production resources under non-stationary demands. A plastic injection molding facility of a medium-size manufacturing firm is studied by a linear programming model. The model results are interpreted and compared to management's actual performance. The linear programming model determines the optimal production schedule and associated levels of inventory and employment for the periods under analysis. The model has been tested and used in production scheduling and inventory control. Analysis of the results has facilitated production planning, decision making, and managerial control.

1. *Introduction*

The optimal use of a firm's production resources is an important management decision function. Optimal production scheduling can often be a critical determinant of profitability in a highly competitive manufacturing environment. Frequently, efficient production planning and scheduling are complicated by fluctuating demands. Seasonal and other variances in product demand are quite common. With such demand variations a manufacturer may require fluctuating levels of production and inventory. The determination of the most efficient level of production and inventory may result in a substantial savings to the manufacturer.

The essence of production planning and scheduling is the determination of an optimal level of production which will meet demand at a minimum cost. Several studies exist for determining the optimal mixture of production strategies under fluctuating demands (Ref. 3, 4, 13, and 14).

The purpose of this chapter is to develop and apply a linear programming model of an actual production/inventory system for the determination of a minimum cost solution for supplying forecasted seasonal requirements in current and future periods. In particular, this study will apply the model to a plastic injection molding facility of a medium-size manufacturing firm. The model determines the optimal schedule for production and associated levels of inventory and employment for future periods. Thus, the model aids decision-making relative to production planning and inventory control.

2. *Data Collection and Adjustments*

The data for this study was collected as part of an extensive research effort involving one of the authors and top corporate management.[1] Data was derived

---

[1]To insure corporate security, data and other information are slightly altered.

from current and past accounting, manufacturing, production control, purchasing, and quality control records for a seventeen month period. Special adjustments were made to determine costs and other parameters when the data was not recorded in a usable format. The list of notation used in this study is shown in Table 1.

Constants or Knowns

c	Total cost.
c_1 (c_2)	Regular (overtime) production labor cost per standard hour.
c_3 (c_4)	Cost of increasing (decreasing) the production rate by one standard hour per day.
c_5 (c_6)	Surplus (shortage) inventory cost per standard hour due to excess over the upper limit (shortage below the lower limit) of the optimal inventory level.
I_i' (I_i'')	Minimum (maximum) inventory allowed in period i.
$\hat{I}_i$ $(\check{I}_i)$	Upper (lower) limit of optimal inventory band in period i.
i	Current period, i = 1,2,...,n.
j	Time periods, j = 1,2,...,i.
M_i	Maximum production rate possible in standard hours per day in period i.
S_i	Demand for standard hours in period i.
t_i (t_i')	Number of regular (overtime) production days available in period i.

Variables or Unknowns

X_i (Y_i)	Number of standard hours produced on regular time (overtime) in period i.
ΔX_i^+ (ΔX_i^-)	Increase or positive change (decrease or negative change) in the production rate in standard hours per day in period i.
z_i^+ (z_i^-)	Excess (shortage) in the ending inventory in standard hours above $\hat{I}_i$ (below $\check{I}_i$) in period i.

TABLE 1. LIST OF NOTATION

Determination of the Unit of Production

The injection molding department under study produces approximately 150 different piece parts. These piece parts are molded on seven injection molding presses with a capacity of 200 or 300 tons each. Because this is a multiproduct, multifacility problem, a reduction of all products to a common unit (machine time) is required. This aggregate unit is necessary to logically and meaningfully express the quantities and the associated costs of producing different parts on similar but separate machines.

The department has a well established standards program, which provides past production efficiency and cost accounting records. Since these records yield an accurate measurement of actual performances, machine time units will be expressed in "standard" hours. The use of standard machine hours allows the aggregation of all products into a single product manufactured on a single machine. The aggregate problem becomes one of producing and inventorying standard hours in order to meet seasonal demands.

Decisions affecting the production of the injection molding department are made monthly. Therefore, throughout this study it is assumed that the length of a time period i is one month.

Development of the Cost Relationships

Different production strategies incur different related costs.[2] Three different production strategies are formulated within the model. These are (i) varying the size of the work force, (ii) working overtime, and (iii) varying the level of inventories. Corresponding to these strategies are six basic cost relationships. These are listed below followed by a more detailed discussion in context of this study.

1. Regular production costs--the direct labor cost of producing standard hours during regularly scheduled shifts.

2. Overtime production costs--the variable direct and variable indirect labor costs of producing standard hours on overtime.

3. Costs of increasing the production rate--the cost of hiring and training new personnel as well as other costs which increase (or decrease) with an increase in the production rate.

4. Costs of decreasing the production rate--the cost of laying off personnel as well as other tangible and intangible costs of decreasing the production rate.

5. Excess inventory carrying costs--the additional carrying costs of inventories in excess of the optimal inventory level for a given period.

---

[2]In this study, standard costs were used only to the extent that they allow average estimates of actual future costs. Also, those costs that varied in response to the changes in manufacturing decisions were considered. For a detailed discussion of cost estimation in aggregate planning and scheduling, see References 6, 11, or 12.

6. Inventory shortage costs--the additional stockout costs for carrying
    inventories less than the optimal inventory level for a given period.

(1,2) *Production Costs*

Regular and overtime production costs were each found to be almost linear.
Thus, the basic production cost in period i is a piecewise linear, convex function as shown in Figure 1. Production below a given level L incurs only regular
direct labor costs (c_1) while production above L incurs the overtime penalty
cost ($c_2 > c_1$).

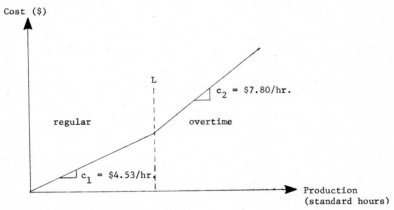

FIGURE 1. AGGREGATE PRODUCTION COST

While injection molding equipment can be run in an automatic mode (not requiring an operator), approximately 96% of all machine time in the production
department is operator assisted. This results in a standard hour of machine time
requiring both an average actual number of hours of machine time and an average
actual input of direct labor hours. On the average, one standard hour of machine
time requires 1.136 hours of actual machine time and 1.316 hours of actual labor
input.

The maximum regular production available from a single machine in an average
day is 21.12 standard hours. The production department has seven machines with
a total maximum capacity of 148 standard hours of production per day. The direct
labor cost (c_1) of one standard hour is $4.53. This direct labor cost is the
direct labor wage divided by the average labor efficiency. The cost of producing
on overtime is greater than 150% (time and a half) of the direct labor wage because of increased indirect labor costs. While the cost of indirect labor is
fixed over the range of regular production (less than L in Figure 1), it is not

fixed for overtime production. The direct and indirect labor premiums are approximated as c_2 = \$7.80 per standard hour (see Figure 1).

(3,4) *Costs of Changes in the Production Rate*

The production rate (X_i/t_i) of a given month is the quotient of the total regular production (X_i) in standard hours divided by the number of days (t_i) in that month. A change in the production rate (ΔX_i) is by definition equal to $(X_i/t_i - X_{i-1}/t_{i-1})$. The production rate occurring on overtime (Y_i/t_i') is also in standard hours per day; it is constrained to be less than or equal to the regular production rate (X_i/t_i). This constraint assures that the maximum number of operators available for overtime is at most the same number of operators as scheduled for regular production.

The primary cost of increasing or decreasing the regular production rate within the maximum of 148 hours per day occurs because of the necessary changes in the size of the work force. The cost of increasing the regular production rate is based on the actual and estimated average costs of hiring and training expenses of new operators, and the estimated cost increases occurring because of added indirect and supporting activities, such as production control, and manufacturing engineering.

The average cost of hiring an operator is approximately \$156.86. On the average, the additional operator produces 6.08 standard hours of output per day. The cost of increasing the production rate by one standard hour per day (c_3) is thus approximately \$25.80 = \$156.86/6.08 (see Figure 2).[3]

The cost of decreasing the regular production rate is based on the actual and estimated average cost of terminating a worker. These costs include unemployment compensation, terminating expenses, production disruption, and an estimate of intangible costs.

The average cost of terminating an operator was determined to be \$134.24 per layoff. A reduction of one operator means a decrease of 6.08 standard hours in the production rate. The cost of decreasing the production by one standard hour per day (c_4) is therefore approximated by \$22.08 = \$134.24/6.08 (see Figure 2).

Figure 2 shows a single linear cost function for increases or decreases in the production rate. Ideally, there should be a separate curve for each starting production rate. The cost of increasing the production rate from 100 hours per day to 106.8 hours per day could be appreciably different than an increase of a like amount near the maximum production rate. However, because major costs are

---

[3]It is assumed that fractional operators can be hired or laid off; by using average costs this assumption is somewhat weakened. Also, random fluctuations in the actual production rate is expected and is not included in the model.

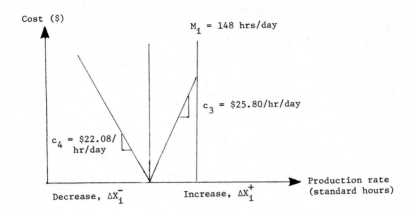

FIGURE 2.   COSTS OF CHANGES IN THE AGGREGATE PRODUCTION RATE

associated with hiring and laying off, the costs of changes in the production
rate are taken as independent of the previous rate.

(5,6) *Inventory Penalty Costs*

Inventory penalty costs are incurred whenever inventories are either higher
or lower than the optimal inventory level for that month. In formulating the
total cost function, it is assumed that the inventory level at the end of each
month is the average inventory level occurring during the month.

Economic lot size formulae are used to determine the optimal production batch
size for each piece part. By adding, for each part, the average safety stock to
one-half the optimal batch size, we obtain its optimal average inventory. Then
by adding together these optimal average inventories for all the parts that are
stocked, an optimal aggregate average inventory for all piece parts is obtained
(Ref. 6). Management felt that an average inventory range is more meaningful
than one inventory figure. As shown in Figure 3, a band of inventory is given
as the optimal ending inventory for any month. This band is found by taking
$\pm 2.5\%$ of the optimal aggregate average inventory.

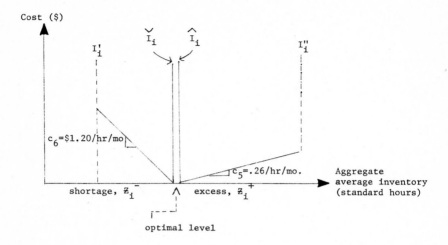

FIGURE 3.   INVENTORY COSTS

When inventories fall below the band, an expense of c_6 = \$1.20 per standard
hour is incurred as a stockout expense. When inventories are in excess of this
band an excess inventory or carrying cost of c_5 = \$.26 per standard hour is in-
curred. The piecewise linear cost function shown in Figure 3 represents an average linear approximation of several total inventory cost functions. Theoretical-
ly, a whole set of curves, one for each level of demand (S_i) is necessary to
describe each inventory cost function. However, such accuracy was not necessary
in this study.

3. *The Model*

The objective of the model is to schedule production and employment so as to
minimize total cost over a given time horizon of n months. The linear programming
model is below.

$$\text{Minimize } c = c_1 \sum_{i=1}^{n} X_i + c_2 \sum_{i=1}^{n} Y_i + c_3 \sum_{i=1}^{n} \Delta X_i^+ + c_4 \sum_{i=1}^{n} \Delta X_i^- +$$

$$c_5 \sum_{i=1}^{n} z_i^+ + c_6 \sum_{i=1}^{n} z_i^-$$

subject to

$$M_i - \frac{X_i}{t_i} \geq 0 \qquad (i = 1,\ldots,n) \qquad (1)$$

$$\frac{X_i}{t_i} - \frac{Y_i}{t_i'} \geq 0 \qquad (i = 1,\ldots\;) \qquad (2)$$

$$\Delta X_i^+ - \frac{X_i}{t_i} + \frac{X_{i-1}}{t_{i-1}} \geq 0 \qquad (i = 1,\ldots,n) \qquad (3)$$

$$\Delta X_i^- + \frac{X_i}{t_i} - \frac{X_{i-1}}{t_{i-1}} \geq 0 \qquad (i = 1,\ldots,n) \qquad (4)$$

$$z_i^+ + \hat{I}_i - \sum_{j=1}^{i} (X_j + Y_j - S_j) \geq 0 \qquad (i = 1,\ldots,n) \qquad (5)$$

$$z_i^- - \check{I}_i + \sum_{j=1}^{i} (X_j + Y_j - S_j) \geq 0 \qquad (i = 1,\ldots,n) \qquad (6)$$

$$\check{I}_i - z_i^- - I_i' \geq 0 \qquad (i = 1,\ldots,n) \qquad (7)$$

$$I_i'' - z_i^+ - \hat{I}_i \geq 0 \qquad (i = 1,\ldots,n) \qquad (8)$$

and

$$X_i, \; Y_i, \; \Delta X_i^+, \; \Delta X_i^-, \; z_i^+, \; z_i^- \geq 0 \quad (i = 1,\ldots,n) \qquad (9)$$

where $\hat{I}_i$, $\check{I}_i$, I_i', I_i'', M_i S_i, t_i, and t_i' are known constants.

In the above model, the production constraint (1) guarantees that the production rate in a period, X_i/t_i, does not exceed the maximum rate M_i. Inequality (1) also defines the maximum number of standard hours available from regular production during period i. The overtime constraint (2) ensures that the production rate available on overtime, Y_i/t_i', does not exceed the rate currently used during regular production. As in inequality (1), inequality (2) limits the maximum number of standard hours available from overtime production to $t_i'(X_i/t_i) \geq Y_i$. Positive production rate changes, ΔX_i^+, are defined greater than or equal to $(X_i/t_i - X_{i-1}/t_{i-1})$ by inequality (3), and negative production rate changes, ΔX_i^-, are defined as being greater than or equal to $-(X_i/t_i - X_{i-1}/t_{i-1})$ by (4). Inequality (5) states that the excessive inventory level z_i^+ above $\hat{I}_i$ is at least as large as the difference between the ending inventory at period i,

$\sum\limits_{j=1}^{i} (X_j + Y_j - S_j)$, and the upper limit of the optimal inventory band, $\hat{I}_i$. For

expression (5) to be valid, the initial inventory in period 1 has to be adjusted
to zero by decreasing the production demand S_1 by the beginning inventory figure.
If the beginning inventory is greater than S_1, the remainder is subtracted from
subsequent demands $(S_2, S_3,$ etc.) until it does equal zero (Ref. 10). Inequality
(6) states that the shortage of ending inventory z_i^{-} below $\check{I}_i$ is the difference
between the lower limit of the optimal inventory band, $\check{I}_i$ and the ending inventory

at period i, $\sum\limits_{j=1}^{i} (X_j + Y_j - S_j)$.

The minimum inventory restriction is given in inequality (7). If the ending
inventory in period i is less than the lower limit of the optimal inventory
band $(z_i^{-} > 0)$, then inequality (7) requires the ending inventory $(\check{I}_i - z_i^{-})$ to
be at least the minimum allowed inventory I_i'. Inequality (8) represents a maximum inventory constraint. Management feels that inventories of a given period
should never be greater than a given level I_i''. When the ending inventory of
period i is greater than the upper limit of the optimal inventory band $(z_i^{+} > 0)$,
this inequality limits the total inventory $(z_i^{+} + \hat{I}_i)$ of any period to be at most
I_i''.

Basic Assumptions

The following assumptions were used in the formulation.
1. Demand for production is given in standard hours as S_i in period i
    $(i = 1,\ldots,n)$.
2. The variable cost of production in period i is a piecewise linear function
    as shown in Figure 1.
3. The cost of changes in the production rate $(\Delta X_i^{+}, \Delta X_i^{-})$, are each linear
    functions as shown in Figure 2.
4. The costs of changes in the production rate $(\Delta X_i = X_i/t_i - X_{i-1}/t_{i-1})$
    are independent of the previous production rate (X_{i-1}/t_{i-1}).
5. The total inventory cost function for all levels of demand S_i is a com-
    mon piecewise linear function as shown in Figure 3.
6. The regular production rate is limited to $M_i = 148$ standard hours per day.
    Additional or decreased capacity through changes in the number of machines
    is not a decision variable.
7. A single production facility is assumed; the facility produces a single
    product.

8. Back orders and lost sales are not allowed.

9. Material costs and other indirect expenses are fixed within the region of study.

10. The model assumes a three-shift operation without the alternative of starting up and shutting down a given shift.

11. Production scheduled for the month is produced within the month.

12. A 17 month time period (January 1972 to May 1973) was used. Note that the optimal solution is usually sensitive to the length of the time horizon

4. Results

The solution of the linear programming model was completed on an IBM 370/135 computer. The computer program used was IBM's Mathematical Programming System/360 program number 360A-CO-14X (Linear and Separable Programming). The problem, as formulated, had 193 constraints and 144 variables. The results for the 17 month period are compared to actual management performance in Figures 4 through 8, and Tables 2 and 3.

Model Production Performance vs. Actual Management Decisions

Management's initial interest in this study was in comparing actual past performances with that of the model. This comparison is the first step in a continued study and application of the model. The model solution provides management with a good basis from which to compare and analyze past and future production policies and procedures.

The results of the model are compared to management's performance fully realizing that many factors may have accounted for the differences. These factors can be classified as either forecasting or policy influences. Similar studies of the past have shown that the majority of penalty costs have resulted from sub-optimal or nonexistent production planning and not from forecasting errors (Ref. 6). As indicated in Table 2, almost all costs of the model is substantially lower than management's performance. Total penalty costs for management's decisions were 3.5 times higher than the model results.

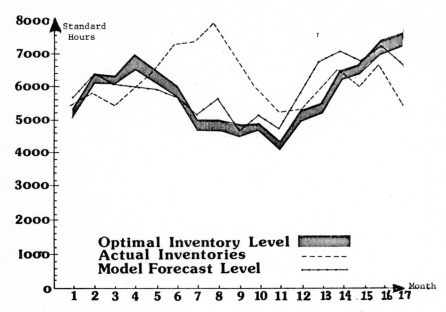

FIGURE 4. ENDING INVENTORY

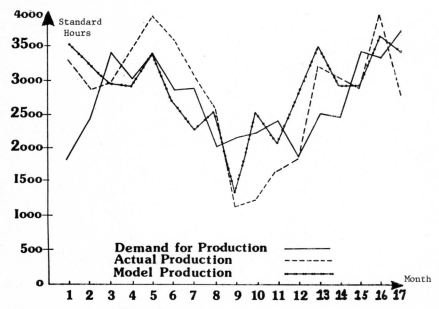

FIGURE 5. AGGREGATE PRODUCTION LEVEL

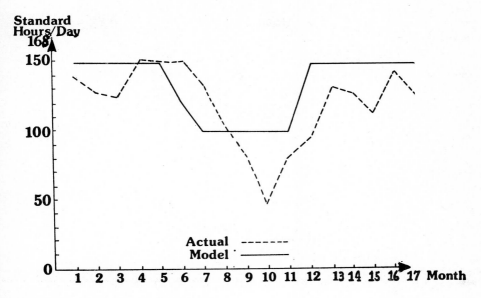

FIGURE 6.   AGGREGATE PRODUCTION RATE

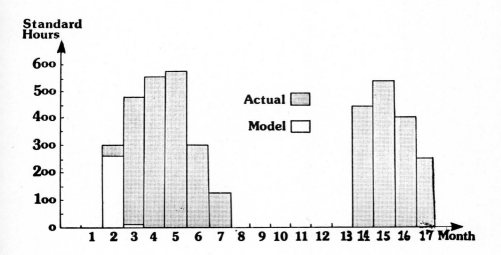

FIGURE 7.   OVERTIME PRODUCTION

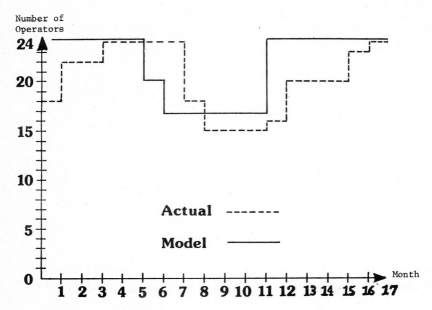

FIGURE 8.   LEVEL OF EMPLOYMENT

COST	MODEL	ACTUAL	PERCENTAGE ACT./MODEL
Changes in the Production Rate	$ 3,373	$ 3,404	101
Excess Inventory	1,103	2,801	254
Inventory Stockout	1,581	5,234	330
Overtime Premium	907	13,140	1,448
Total Penalty	6,964	24,580	353
Direct Labor	216,198*	216,198	100
Total	$223,162	$240,778	107.9

*The model's direct labor cost actually equaled $221,739; the difference in cost being the model's ending inventory of 6,743 standard hours versus the actual ending inventory of 5,520 standard hours. For cost comparisons the model was not penalized for anticipating future demands and producing this higher ending inventory.

TABLE 2.   COMPARISON OF TOTAL COSTS

COSTS	MODEL	ACTUAL
Changes in the Production Rate	48.4	13.8
Excess Inventory	15.8	11.4
Inventory Stockout	22.7	21.3
Overtime Premium	13.1	53.5
Total Penalty	100.0	100.0

TABLE 3. COMPARISON OF PENALTY COSTS (in percents)

Management depended upon overtime as the predominant mode in responding to fluctuations in demand. Yet, the model used overtime as the least dominant mode. Actual overtime penalty cost for management was about 14 times higher than the model indicated. The model incurred overtime penalty costs in the second and third months of the first year. These costs would have certainly been avoided had several prior months been available for anticipating the upcoming seasonal peak.

A quick analysis of the cost coefficients shows that one hour of overtime premium ($3.27 = $7.80 - $4.53) incurs the same penalty cost of approximately 12.5 (or 3.27/.26) months of carrying inventory. Under certainty, overtime expense will not be incurred unless the capacity to build up inventories does not exist within any of the preceding 12 months.

Changes in the production rate are the predominant expense of the model. Only three changes occur throughout the 17 month period. Two decreases occurred in the sixth and seventh months of the first year and one increase to the maximum production rate occurred in the twelfth month. Actual management performance increased and decreased production slightly more frequently than the model. Quite evident is the fact that actual increases and decreases lag the model's decision by a month or more. While such lags might occur with imperfect forecasting, they do create de-stabilizing effects upon future decisions.

In the first year, it is evident that production rates were not decreased as rapidly as in the optimal solution. As shown in Figure 6, maximum rates were maintained into the sixth and seventh months resulting in the excessive inventory (Fig. 4) and overtime costs during these months. These same inventories were depleted because of the low production rates in the ninth through twelfth months. As shown in Table 2, the actual and model costs for changes in the level of production are identical.

The model maintained inventories closer to optimal levels than did actual decisions. Those periods in which the model did deviate from the optimal inventory

are significant. They exemplify the model's decision to incur excessive inventory
and stockout costs rather than use overtime during peak periods. Quite logically,
the model stockout costs are incurred during seasonal peak periods and excessive
inventory costs are incurred prior to these peaks. Actual management decisions
resulted in a stockout cost which was approximately 3.3 times higher than that of
the model. These stockout costs result from delays in increasing the production
rates as discussed above.

Unfortunately, excessive inventories occurred during periods in which they
were unnecessary. Inventories were at their highest during the sixth through the
ninth months. The inventory level of the tenth month was quite favorable relative
to approaching seasonal peaks. However, for several reasons, management failed
to maintain its favorable inventory position from the tenth month through the
fifth month of the following year (or month 17).

References

1. Beebe, J. H., C.S. Beightler, and J. P. Stark; "Stochastic Optimization
    of Production Planning," Operations Research 16, 799-818 (1968).

2. Buffa, E. A., and W. H. Taubert; Production-Inventory Systems: Planning and
    Control, Richard D. Irwin, Inc., Homewood, Ill., 1972.

3. Crowston, W. B., W. H. Hausman, and W. R. Kampe II; "Multistage Production
    for Stochastic Seasonal Demand," Management Science 19, 924-935 (1973).

4. d'Epenous, F.; "A Probabilistic Production and Inventory Problem," Management
    Science 10, 98-108 (1963).

5. Hansmann, F.; Operations Research in Production and Inventory Control, John
    Wiley and Sons, Inc., New York, 1962.

6. Holt, C. C., F. Modigliani, J. F. Muth, and H. A. Simon; Production Planning,
    Inventories, and Work Force, Prentice-Hall, Inc., Englewood Cliffs, N. J.,
    1960.

7. I.B.M. Corporation, Mathematical Programming System/360 Version 2: Linear
    and Separable Programming -- User's Manual (GH20-0976-2) I.B.M. Corporation,
    White Plains, N. Y., 1971.

8. Kortanek, K. O., D. Sodar, and A. L. Soyster; "Multi-product Production
    Scheduling via Extreme Point Properties of Linear Programming," Naval
    Research Logistics Quarterly 15, 287-300 (1968).

9. Love, S. F.; "Dynamic Deterministic Production and Inventory Models with
    Peicewise Concave Costs," Technical Report #3, Department of Operations
    Research, Stanford University, January 1968.

10. Magee, J. F.; Production Planning and Inventory Control, McGraw-Hill Book
    Company, New York, 1958.

11. McGarrah, R. E.; Production and Logistics Management: Text and Cases, John
    Wiley and Sons, Inc., New York, 1963.

12. Orr, D.; "A Random Walk Production-Inventory Policy: Rational and Implementation," Management Science 9, 108-122 (1962).

13. Silver, E. A.; "A Tutorial on Production Smoothing and Work Force Balancing," Operations Research 15, 985-1010 (1967).

14. Veinott, Jr., A.F.; "Optimal Policy in a Dynamic, Single Product, Non-stationary Inventory Model with Several Demand Classes," Operations Research 13, 761-778 (1965).

15. Vergin, R. C.; "Production Scheduling Under Seasonal Demand," Journal of Industrial Engineering 17, 260-266 (1966).

16. Winters, P. R.; "Constrained Rules for Production Smoothing," Management Science 8, 470-481 (1962).

17. Zangwill, W. I.; "A Deterministic Multiproduct Multifacility Production and Inventory Model," Operations Research 14, 486-507 (1966).

Chapter 6

ON MINIMIZING WORKMEN'S COMPENSATION PAYMENT
SCHEDULES BY LINEAR PROGRAMMING

Harvey M. Salkin
Department of Operations Research
Case Western Reserve University
Cleveland, Ohio 44106

Burton V. Dean
Department of Operations Research
Case Western Reserve University
Cleveland, Ohio 44106

Chien H. Lin
Department of Operations Research
Case Western Reserve University
Cleveland, Ohio 44106

and

Susumu Morito
Department of Operations Research
Case Western Reserve University
Cleveland, Ohio 44106

Abstract. Workmen's compensation usually relates
to a fund established by the employer to protect
its employees in the event of certain types of unemployment. In the United States, the creation
and maintenance of workmen compensation funds is
usually guided by state legislation. Depending
on the employers location, one of three methods
is usually used to create and maintain the compensation fund; namely, the employer may have private insurance, be self-insured (via a pool of its
own money), or have insurance from a state fund.
In the last case, the amount of payment that the
employer makes to the state fund is dependent
upon how the company's subsidiaries are treated
by the accounting organization. In particular,
each may contribute separately, all together, or
in various aggregated forms. It turns out that
an integer programming model can be used to determine the best way for a corporation's subsidiaries to contribute to the fund so that the total contribution is minimized. The integer program is often solved using linear programming.
We present the model and computational results
reflecting 1974 data from several Ohio companies.

1. *Introduction*

Workmen's compensation usually relates to a fund established by the employer
to protect its employees in the event of certain types of unemployment. In the
United States, the creation and maintenance of workmen compensation funds is
usually guided by state legislation. Depending on the employers location, one of
three methods is usually used to create and maintain the compensation fund; namely, the employer may have private insurance, be self-insured (via a pool of its
own money), or have insurance from a state fund. In the last case, the amount
of payment that the employer makes to the state fund is dependent upon how the
company's subsidiaries are treated by the accounting organization. In particular, each may contribute separately, all together, or in various aggregated
forms. For example, Figure 1 displays a corporation with 3 subsidiaries. The
corporation has the option of contributing to the state fund by considering itself to be 3 companies (Figure 2a), 2 companies (Figure 2b), or 1 company (Figure
2c). In each case, the annual payment to the state fund will be different. As a
modest amount of interest is given, in dollars sent to the fund, and this money
is sorely needed for working capital, the corporation desires to pay as little as
possible. Therefore, subject to the various restrictions and guidelines, the
problem is to determine how the subsidiaries should be aggregated so that the
corporation's total annual contribution is minimized. This problem can be formulated as an integer program which can usually be solved using linear programming.

We present the details using the state of Ohio and its particular computations for determining payments to the workmens compensation state fund. Computational results reflecting several Ohio firms are then given.

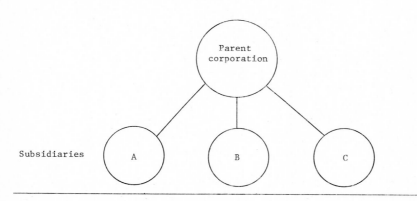

FIGURE 1:   A CORPORATION WITH 3 SUBSIDIARIES

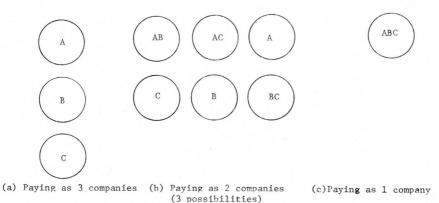

(a) Paying as 3 companies   (b) Paying as 2 companies   (c)Paying as 1 company
                                  (3 possibilities)

FIGURE 2:   POSSIBLE CONFIGURATIONS FOR A 3 SUBSIDIARY CORPORATION

2. *The Problem*

In Ohio, the employers (i.e., corporations, companies, partnerships, etc.) are required to be self-insured or insured in an exclusive state fund. The former requires that the employer gives evidence of general financial responsibility and posting a large bond. The latter requires that the employer pays an annual contribution to the state fund maintained by the Workmen's Compensation Commission. The Commission also has the responsibility of classifying industries and establishing the annual premium rate sufficiently large to provide an adequate fund to pay-out the compensation incurred each year. Prior to the premium rate computation date, the administrator of the Commission determines the benefits paid out of each employer's account and the employer's contribution. The administrator will then determine the positive or negative balance of each account by calculating the difference of such contributions and interest over the benefits distributed. The premium rate is then determined by computing the ratio of the positive or negative balance to the employer's average annual payroll. The contribution due for the next year is then equal to the premium rate, determined from a given table, using the ratio (see Table 1), multiplied by the expected annual payroll. More specifically, let

R = the balance in the reserve fund

P = the average annual payroll

RT = (R/P) x 100%   (ratio)

PT = the premium rate = $f(RT)/100 + .003$[1]

and EP = the expected annual payroll,

then the total annual contribution due = EP x PT. In Ohio, the 1974 premium rates for one particular contribution schedule is given in Table 1.

Example:

R = \$442,360.85

P = \$4,533,643.52

EP = \$5,900,000.00

RT = R/P = 9.76%

PT = f(RT)/100 + .003 = 0.8%

Total annual contribution = EP x PT = \$47,200.00

3. *The Model*

As previously mentioned, the law also permits the establishment of joint accounts by a group of employers in a single firm, and is especially important if the employers represent subsidiaries of a parent corporation wishing to minimize

---

[1]$f(RT)$ is given in Table 1 and .003 is a premium surcharge.

RT	f(RT)	RT	f(RT)
RT < -1.0%	3.7%	6.0% ≤ RT < 6.5%	1.5%
-1.0% ≤ RT < 0.0%	3.6%	6.5% ≤ RT < 7.0%	1.3%
0.0% ≤ RT < 1.0%	3.5%	7.0% ≤ RT < 7.5%	1.1%
1.0% ≤ RT < 1.5%	3.4%	7.5% ≤ RT < 8.0%	0.9%
1.5% ≤ RT < 2.0%	3.3%	8.0% ≤ RT < 8.5%	0.8%
2.0% ≤ RT < 2.5%	3.1%	8.5% ≤ RT < 9.0%	0.7%
2.5% ≤ RT < 3.0%	2.9%	9.0% ≤ RT < 9.5%	0.6%
3.0% ≤ RT < 3.5%	2.7%	9.5% ≤ RT < 10.0%	0.5%
3.5% ≤ RT < 4.0%	2.5%	10.0% ≤ RT < 10.5%	0.4%
4.0% ≤ RT < 4.5%	2.3%	10.5% ≤ RT < 11.0%	0.3%
4.5% ≤ RT < 5.0%	2.1%	11.0% ≤ RT < 11.5%	0.2%
5.0% ≤ RT < 5.5%	1.9%	11.5% ≤ RT < 12.0%	0.1%
5.5% ≤ RT < 6.0%	1.7%	12.0% ≤ RT	0.0%

TABLE 1.

its total contribution. The calculation of the joint contribution is the same as the calculation of an individual account computed from an aggregate average annual payroll and reserve fund. The question is: How should the aggregation be determined so that each subsidiary is in one and only one account, and the total annual contribution is minimized? This problem can be formulated as an integer linear program as follows: Let the subsidiaries be numbered as $\{1,2,\ldots m\}$, and let $S = \{S_1,\ldots,S_n; n = 2^m - 1\}$ be the set of all possible combinations of the m employers. Corresponding to each $S_j \varepsilon S$, there is an m x 1 vector a_j with components $a_{ij} = 1$ or 0 depending on whether the ith subsidiary is in S_j ($a_{ij} = 1$) or not ($a_{ij} = 0$). Associated with each S_j is a contribution or cost c_j (i.e., c_j = EP x PT). Then the problem is modelled as the problem below.

(SP)   Minimize $\displaystyle\sum_{j=1}^{n} c_j x_j$

subject to $\displaystyle\sum_{j=1}^{n} a_j x_j = e$

and $x_j = 0$ or 1 $(j=1,\ldots,n)$,

where e is an m column of ones, $a_j = (a_{ij})$ is an m binary column, and $n = 2^m - 1$.

An Example   (Problem 1: Corporation A)

To see that model SP represents our problem, consider the 4 subsidiary problem whose data is below.

Subsidiary	Average Payroll(P)	Reserve Fund(R)	Estimated 1975 Payroll(EP)
1	$17,997.54	$8,300.13	$6,000
2	405,555.48	4,598.76	400,000
3	5,918,533.41	390,298.51	6,500,000
4	413,542.66	16,202.79	400,000

As these are 4 subsidiaries, $m = 4$ and $2^m - 1 = 15$, meaning that 15 binary columns can be generated corresponding to all the possible aggregation schemes. They are listed below. Notice that (e.g.) a_6 means that subsidiary 1 and 3 will

Variable Subsidiary	x_1 a_1	x_2 a_2	x_3 a_3	x_4 a_4	x_5 a_5	x_6 a_6	x_7 a_7	x_8 a_8	x_9 a_9	x_{10} a_{10}	x_{11} a_{11}	x_{12} a_{12}	x_{13} a_{13}	x_{14} a_{14}	x_{15} a_{15}	e
1	1	0	0	0	1	1	1	0	0	0	0	1	1	1	0	1	=1
2	0	1	0	0	1	0	0	1	1	0	1	1	1	0	1	1	=1
3	0	0	1	0	0	1	0	1	0	1	0	1	1	0	1	1	=1
4	0	0	0	1	0	0	1	0	1	1	1	0	1	1	1	1	=1

be aggregrated if $x_6 = 1$, and the fact that each $x_j = 0$ or 1 with each row summing to 1 guarantees each subsidiary is in one and only one aggregation.

For each column a_j, its cost c_j is computed using c_j = EP x PT, where PT = $f(RT)/100$ + .003 and RT = (R/P) x 100%. The terms PT, RT, and EP may, of course, represent aggregated figures. For example, the calculation for c_6 gives

R = 8,300.13 + 390,298.51 = 398,598.64

P = 17,997.54 + 5,918,533.41 = 5,936,530.95

RT = (R/P) x 100% = 6.71%

PT = f(RT)/100 + .003 = .013 + .003 = .016

EP = 6,000 + 6,500,000 = 6,506,000

c_6 = EP x PT = $104,096.

Similar calculating gives the remaining payment figures.

c_j →	1	2	3	4	5	6	7
j →	18	14,800	104,000	11,200	8,120	104,096	8,120

$c_j \rightarrow$	8	9	10	11	12	13	14	15
$j \rightarrow$	124,200	25,600	124,200	124,308	24,180	110,496	131,400	131,508

Replacing the x_j = 0 or 1 constraints by $0 \leqslant x_j \leqslant 1$ (j=1,...,15) produces a linear program.[2] Its optimal solution is $x_2 = x_{13}$ = 1 and x_j = 0 otherwise, with a minimal total payment of \$125,296. This means that subsidiary 2 should pay separately, while 1, 3, and 4 should make a combined payment.

4. *Computational Experience.*

Optimal aggregation configurations were computed for several Ohio firms. In each problem, model SP was constructed using a subroutine whose inputs are the number of subsidiaries (m), and for each, the average payroll (P), the reserve fund (R), and the expected payroll (EP). Naturally, a rate schedule, such as Table 1, is also required. The resulting model was then solved as a linear program by substituting $x_j \geqslant 0$ for x_j = 0 or 1 (j=1,...,n). Each run required less than a second using a UNIVAC 1108 computer and the SCA4 code.[3] Every optimal solution turned out to have only integer values for the variables. Of course, had this not been the case an integer programming code should be used.[4] Some data and results are tabulated starting on the next page.

5. *Conclusion*

The situation we have presented in this chapter can easily be formulated as model SP and, in most cases, solved as a linear program. This procedure is obviously quite effective and, moreover, the model can simply be extended to accomodate similar scenarios and other constraints. However, as m, the number of subsidiaries, increases the number of possible aggregations (or variables) increases dramatically. For example, when m = 30, $n > 10^9$. To avoid the time, effort, and money required to solve large scale linear programs, heuristics can be used to reduce the number of subsidiaries prior to formulating model SP. In particular, certain subsidiaries can sometimes be aggregated initially so that the value of

---

[2]In computations, the upper bound constraints $x_j \leqslant 1$ (j=1,...,n) are omitted since the zero-one coefficient matrix, the right hand side of ones, and the positive cost values guarantee that no variable will exceed 1.

[3]This code, developed by H. Salkin, is a variation of the one described in [2]. It also contains a special purpose integer programming algorithm and can thus enforce the integrality conditions if necessary.

[4]Model SP is a special type of integer program called a set partitioning problem. These are extensively discussed in [4] and [5]. A similar application is in [1].

Problem 2: Corporation B

Input...m = 8

Subsidiaries	Average Payroll	Reserve Fund	Estimated 1975 Payroll
1	96,889.59	8,924.34	118,000
2	781,150.56	100,051.72	1,050,000
3	4,200.00	418.54	4,200
4	4,783.33	869.37	4,200
5	4,200.00	207.34	4,200
6	165,248.55	12,444.84	228,000
7	5,496,481.66	473,178.75	7,500,000
8	7,459,980.81	580,134.64	9,000,000

Output...

Aggregation	Average Payroll	Reserve Fund	(R/P)x100%	RP	Estimated Payroll	Payment
1,2,5,6	1,047,489	121,628	11.61	0.40	1,400,200	5,601
3,4	8,983	1,288	14.34	0.30	8,400	25
7,8	12,956,462	1,053,313	8.13	1.10	16,500,000	181,500
					Total Payment	$187,126

Problem 3: Corporation C

Input...m = 5

Subsidiaries	Average Payroll	Reserve Fund	Estimated 1975 Payroll
1	336,273.23	39,299,20	375,000
2	544,908.79	47,751.53	550,000
3	41,855.75	3,953.31	55,000
4	24,717.26	433.32	40,000
5	70,752.08	4,985.02	65,000

Output...

Aggregation	Average Payroll	Reserve Fund	(R/P)x100%	RP	Estimated Payroll	Payment
1,4	360,990	39,733	11.01	0.50	415,000	2,075
2,5	615,661	52,737	8.57	1.00	615,000	6,150
3	41,856	3,953	9.45	0.90	55,000	495
					Total Payment	$8,720

1.   Current account overdrafts.

2.   Discount of trade acceptances.

3.   Borrowing against certificate of indebtedness.
     (Schuldscheindarlehen).

4.   Advance financing.   (1)

5.   Loans in foreign currencies.

6.   Credit balances in current account.

7.   Time deposits.

Hereby 1 to 5 represent credit extensions from banks, and 6 and 7 deposits placed
with banks. For reasons discussed later we divide our planning period (e.g., a
fiscal year) into N decades (periods of ten days). We admit availments (using)
and deposits of credits only at the beginnings of decades numbered $0,\ldots,N$.

By $X_{i,k,j,m}$ we denote the amount of the kth type of credit ($k=1,\ldots,7$) in the
i^{th} decade ($i=1,\ldots,N$) obtained from or deposited on the jth bank ($j=1,\ldots,J(k)$)
with the mth duration (i.e., some fixed time period) ($m=1,\ldots,M(j,k)$); where $J(k)$
is the number of banks from which the kth type credit can be obtained and $M(j,k)$
is the total number of durations for which the kth credit can be obtained from or
deposited on the jth bank. In correspondence $Y_{i,k,j,m}$ means the credit amounts
repayable ($k=1,\ldots,5$) and the anticipated credit advices ($k=6,7$), respectively.
By definition, we have

$$X_{i,k,j,m} \geq 0 \quad \text{and} \quad Y_{i,k,j,m} \geq 0. \qquad (2)$$

By $L_{k,j,m}$ we denote the number of decades in the mth duration of the kth type of
credit from the jth bank; that is, money has to be repaid at the end of $L_{k,j,m}$
decades.

Credits 1 and 6 are regarded to have a duration of one decade which means
$M(j,1) = M(j,6) = 1$ with $L_{1,j,1} = L_{6,j,1} = 1$. Credit 2 is also restricted to
have a fixed duration (in our case 7 decades) which means $M(j,2) = 1$ (i.e., one
duration) with $L_{2,j,1} = 7$ (i.e., the length of the duration is 7 decades). For
the remaining credits we have several durations which means $M(j,k) \geq 1$ for
$k=3,4,5,7$ and $j=1,\ldots,J(k)$. (In reality there are durations of 3, 6, 9, 18, and
36 decades.)

For the beginning of the planning period

$$Y_{i,k,j,m} \qquad (i = 0,1,\ldots,L_{k,j,m}-1) \qquad (3)$$

anticipated credit advices due to banks are known from earlier periods. Since
the credits have to be repaid at the end of their durations, we have

$$Y_{i,k,j,m} = X_{i-L_{k,j,m},k,j,m} \qquad (\mathcal{1} = L_{k,j,m},\ldots,N). \qquad (4)$$

Thus, by (4), for $i \geq L_{k,j,m}$ the variables $Y_{i,k,j,m}$ may be eliminated from the model. Further, all credits must be repaid during the planning period, or

$$X_{i,k,j,m} = 0 \qquad (i = N - L_{k,j,m}+1,\ldots,N). \qquad (5)$$

An essential practical requirement for the variables of the credit type $k=3,4,5,7$ is that

$$X_{i,k,j,m} \quad \text{integer} \quad (k=3,4,5,7). \qquad (6)$$

As required by the banks the credits for $k=3,4,7$ are only available as integer multiples of one million DM. For $k=5$ there is a further complication; in particular, the integer condition refers to the currency units (e.g., 100,000 United States dollars).

For the same credits the banks further stipulate time independent minimum and maximum ranges of credit availments; thus, for $k=3,4,5,7$ and for $i=0,1,\ldots,N-L_{k,j,m}$

$$P_{k,j,m} \leq X_{i,k,j,m} \leq Q_{k,j,m} \quad \text{or} \quad X_{i,k,j,m} = 0. \qquad (7)$$

Here, the $P_{k,j,m}$ and the $Q_{k,j,m}$ are integers which for $k=5$ are understood to be in foreign currency units.

A last restriction for credit availments is in the $k=2$ (trade acceptances) category. In particular, only a certain portion p ($0 < p \leq 1$) of the estimated accounts payable in the ith period, denoted by

$$B_i \qquad (i = 0,1,\ldots,N-1) \qquad (8)$$

accumulated for each duration (or time period) of $L_{2,j,1}$ decades, can be paid via trade acceptances. Using our notation these constraints are

$$\sum_{h=h(i)}^{i} \sum_{j=1}^{J(2)} X_{h,2,j,1} \leq P \sum_{h=h(i)}^{i} B_h \qquad (i = 0,1,\ldots,N-1), \qquad (9)$$

with

$$h(i) = \text{maximum} \ (0,i-L_{2,j,1}+1).$$

Here, the right hand side term in (9) represents the accumulated accounts payable over $L_{2,j,1}$ decades at period i multiplied by the factor p. The left hand side is the total credit type 2 obtainable from all the banks in the time period from h(i) to i. Also, on the left hand side in (9) certain $X_{i,k,j,m}$ as specified in (5), must be set to zero.

In practice, the banks set limits for the existing credits which must not be exceeded. The limits for seasonal credits may also be time dependent. The credit limits mean that the sum of availments minus repayments, i.e., the actual amount, accumulated up to the decade i is restricted by a given upper limit $S_{i,k,j}$, where i denotes the time, k the credit type, and j the bank. These constraints are

$$0 \leq \sum_{m=1}^{M(j,k)} \sum_{h=0}^{i} (X_{h,k,j,m} - Y_{h,k,j,m}) \leq S_{i,k,j} \tag{10}$$

for k=1,...,7 and j=1,2,...,J(k). Because of M(j,1) = M(j,2) = M(j,6) = 1 the first summation in (10) can be dropped for credit types k=1 and 6. The left part of the inequalities (10) stems from the fact that we demand a nonnegative amount for each credit. If there is no limit, e.g., for k=6 in practice, the extreme right hand side of the inequality is dropped.

Using (10) with $Y_{i,k,j,m} = 0$ for $i \geq L_{k,j,m}$ (see (4)) we have for any period i_o $(1 \leq i_o \leq N-1)$

$$S_{i_o,k,j} = 0 \quad \text{implies} \quad X_{i,k,j,m} = 0, \tag{11}$$

for m=1,...,M(j,k) and $i=i_o-L_{k,j,m}+1,...,i_o$.

For k=1 and 6 there is an additional constraint, namely

$$\sum_{h=0}^{i_o+1} (X_{h,k,j,1} - Y_{h,k,j,1}) = 0, \tag{12}$$

whenever $S_{i_o,k,j} \neq 0$ and $S_{i_o+1,k,j} = 0$ since the amount accumulated at decade i_o+1 must be zero. This means that seasonal overdrafts for k=1 and 6 have to be repaid at least at the beginning of the first decade following the season.

As the variables $Y_{i,k,j,m}$ for $i=0,1,...,L_{k,j,m}-1$ are known (see (3)), they may be substituted into (1) and after rearranging the terms we have that

$$0 \leq \sum_{m=1}^{M(j,k)} \sum_{h=0}^{i} X_{h,k,j,m} \leq \hat{S}_{i,k,j} \qquad (i = 0,1,...,L_{k,j,m}-1), \tag{13}$$

where

$$\hat{S}_{i,k,j} = \text{maximum} \left(0, S_{i,k,j} - \sum_{m=1}^{M(j,k)} \sum_{h=i+1}^{L_{k,j,m}-1} Y_{h,k,j,m} \right).$$

When $\hat{S}_{i,k,j} = 0$ the inequality (13) may be dropped.

In addition, certain banks establish a combined credit limit for the k=1 and k=2 credit type. In this case, for k=1, (10) and (12) are valid, but inequality (10) does not apply for k=2. Instead we have the condition that

$$0 \le \sum_{k=1}^{2} \sum_{h=0}^{i} (X_{h,k,j,1} - Y_{h,k,j,1}) \le S_{ij}^{*}, \tag{14}$$

where S_{ij}^{*} is the common limit. For k=2 and $h=0,1,\ldots,L_{2,j,m}-1$ we may use (3) to substitute for $Y_{h,k,j,1}$ in (14).

Now having introduced the credit restrictions let us proceed to the condition of liquidity. The estimated disbursements and receipts, denoted by

$$a_i \quad \text{and} \quad e_i \qquad (i = 0,1,\ldots,N-1), \tag{15}$$

respectively, are known. The estimated accounts payable B_i satisfy $B_i \le a_i$ $(i=0,\ldots,N-1)$. The concern is forced to be liquid at the beginning of each decade; equivalently, the sum of the differences between availments and repayments must always be at least as large as the difference between the disbursements and receipts. Therefore, we have

$$\sum_{k=1}^{7} \sum_{j=1}^{J(k)} \sum_{m=1}^{M(j,k)} f_k (X_{i,k,j,m} - Y_{i,k,j,m}) \ge a_i - e_i \qquad (i=0,1,\ldots,N-1). \tag{16}$$

In (16), f_k is set to 1 for $k=1,\ldots,5$ (availments) and $f_k=-1$ for k=6,7 (deposits). Further, the given amounts (3) as well as the relations (4) and (5) should be used in (16).

Before establishing the objective function we discuss the dividing of the planning period into decades. To be realistic we ought to have used subintervals of a day's length. If we choose approximately one year as the total length of a planning period a huge linear program would result. Moreover, the finance manager would be confronted with an enormous number of single proposals for his activities and therefore may not have time to choose a course of action in each interval.

The objective function can be written as

$$\text{minimize} \quad \sum_{k=1}^{7} f_k \, V_k, \tag{17}$$

where the factors f_k are defined according to (16), and where V_k is the sum of the interest over all credits of type k computed using one or more durations and all N decades. Each term making up V_k is a product of an effective interest factor and of the corresponding availment. The effective interest factor is a sum of the interest rate itself and any other costs, all computed on an annual basis (36 decades). Now we develop the formula for the various V_k.

Let $\gamma_{k,j,m}$ denote the annual interest rate (in percent) for the availment j of credit type k with duration m. We drop the duration index m if there is only one duration. For k=1 the interest is paid and for k=6 the interest is received each quarter year. This effect of compounded interest is considered in calculating the effective availment at the time i. Let $d_{i,h}$ denote the number of decades contained in the hth quarter year of the planning period and following the decade i-1. Let H be the total number of quarter years covering the planning period. Then we have

$$V_k = \sum_{j=1}^{J(k)} \sum_{i=0}^{N-1} \left[\prod_{h=1}^{H} \left(1 + \left(\frac{\gamma_{kj}}{100}\right)\left(\frac{d_{i,h}}{36}\right)\right) - 1\right] (X_{i,k,j,1} - Y_{i,k,j,1}) \quad (k=1,6). \tag{18}$$

Here the term in square brackets represents the compound interest effect and the other factor represents the net amount of money borrowed (k=1) or on deposit (k=6).

For credit type k=2 the total amount of interest in the duration, denoted by $\gamma'_{2,j}$, becoming due at the beginning of the duration, is composed of a bank independent bill stamp tax T (as a percent, for the total duration) and of the annual interest rate $\gamma_{2,j}$. This means that

$$V_2 = \sum_{j=1}^{J(2)} \sum_{i=0}^{N-1} (1 + \gamma'_{2,j}) \gamma'_{2,j} X_{i,2,j,1}, \tag{19}$$

with

$$\gamma'_{2,j} = \frac{L_{2,j,1} \, \gamma_{2,j}}{100} + \frac{T}{100}. \tag{20}$$

In (19), the term $(1 + \gamma'_{2,j})\gamma'_{2,j}$ represents the compound interest effect and $L_{2,j,1} = 7$. Note that by the discussion preceding expression (14) $Y_{i,2,j,1} = 0$ so only $X_{i,2,j,1}$ appears in (19). A similar remark is valid for expressions (21) and (24).

For credit type k=3 the discounted interest $A_{3,j,m}$ (in percent) is for the whole duration. On the other hand, for time deposits (k=4,7) the broker commission $A_{k,j,m}$ (in percent) is on a per annum basis. In both cases the normal interest has to be paid when the principal is due whereas the broker's fee is due at once. Thus we have

$$V_k = \sum_{j=1}^{J(k)} \sum_{i=0}^{N-1} \sum_{m=1}^{M(j,k)} \left((1+A'_{i,k,j,m})A'_{i,k,j,m} + \frac{\gamma_{k,j,m}}{100} \frac{t_i(L_{k,j,m})}{360} \right) X_{i,k,j,m} \qquad (k=3,4,7) \tag{21}$$

with

$$A'_{i,k,j,m} = \frac{f_{ik} A_{k,j,m}}{100} \tag{22}$$

where

$$f_{ik} = \begin{cases} 1 & \text{for } k=3 \\[2mm] \dfrac{t_i(L_{k,j,m})}{360} & \text{for } k=4,7. \end{cases} \tag{23}$$

Here the number $t_i(L_{k,j,m})$ is the duration expressed in days. The subscript i indicates that the duration in days also depends on the position of the decade in the calendar month when the English or French interest computation method is applied. If the English interest computation method is used, the number 360 is replaced by 365.

These last remarks are also valid for credits in foreign currencies (k=5). However, we further take into consideration the buying rate $E_{j,m}$ and the estimated selling rate $F_{j,m}$ (both in currency units). The expression for V_5 is

$$V_5 = \sum_{j=1}^{J(5)} \sum_{i=0}^{N-1} \sum_{m=1}^{M(j,5)} \left((1+A'_{i,5,j,m})A'_{i,5,j,m} + \frac{\gamma_{5,j,m}}{100} \frac{t_i(L_{5,j,m})}{360} + \frac{F_{j,m}-E_{j,m}}{E_{j,m}} \right) X_{i,5,j,m} \tag{24}$$

with

$$A'_{i,5,j,m} = \frac{t_i(L_{5,j,m})}{360} \frac{A_{5,j,m}}{100} . \tag{25}$$

Here $A_{5,j,m}$ is the per annum broker commission in percent. The term $(F_{j,m} - E_{j,m})/E_{j,m}$ is the expected percentage gain or loss. In both (21) and (24) the term in the large paranthesis represents the total compounded interest.

The linear program for short term liquidity management is represented by the objective function (17) and the constraints (5), (7), (9), (10), (11), (12), (13),

(14), and (16), plus the nonnegativity conditions (2). If the integer requirements (6) are added the problem is of the mixed integer type.

3. *Implementation*

In its present form the linear program with (6) is of the mixed integer type.
Because of the large number of integer variables (e.g., about 1,000 in the case of
a planning period of one year and about 28 credits of the sort k=3,4,5,7) the
mixed integer program probably cannot be solved using the known methods and available computer programming packages within a reasonable amount of time. However,
as the credit types k=3,4,5 normally carry lower interest rates than current
account overdrafts (k=1), and credit type k=7 yields higher interest than current
account balances (k=6), it can be expected that the upper integer limits $Q_{k,j,m}$
in (7) are very often attained. Thus, in the actual program we have dropped the
integer conditions (6) and have replaced them by

$$0 \le X_{i,j,k,m} \le Q_{k,j,m} \qquad (k=3,4,5,7). \qquad (26)$$

In most cases this procedure yielded integer values for the $X_{i,k,j,m}$ variables.
If a number of noninteger values appear a heuristic algorithm might be added.

Before describing the computer programming system we illustrate the structure
of the linear program's coefficient matrix using an example. For a planning
period with 21 decades (N=21) we consider the following loan compositions (each
credit at only one bank): one credit k=1, one credit k=1 and one credit k=2
bearing a combined upper limit (see (14)), one further credit k=2, three advances
(k=4) again bearing a combined upper limit with durations of 3, 9, and 18 decades,
one deposit in current account k=6, and one time deposit with a duration of 9
decades. In Figure 1 the maximum N=21 variables for each credit alternative
must be examined in the indicated sequence from left to right. In the diagram a
variable is always represented by a vertical line. The horizonal lines represent
the constraints and the objective function. The shadowed areas mean that the inside crossing points of two straight lines correspond to a matrix coefficient with
the value +1. This also applies if a continuously drawn diagonal intersects such
a crossing point. A dotted diagonal represents coefficients with the value -1.
The rest of the intersections represent zeroes. Only in the objective function
(17) and on the right hand sides are these nonzero coefficients different from ± 1.
We have omitted the availment restrictions (26) for k=4 in the figure because
using the simplex method they are replaced by bounded variables.

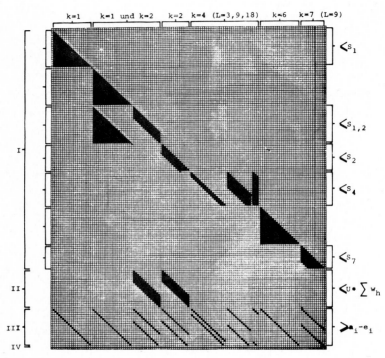

FIGURE 1. A SAMPLE PROBLEM

The organization of the computer programming system appears in Figure 2. We have not shown how the master file is established. Its input are credit alternatives characterized by the bank name, the credit type, the duration, the interest rate, the fee or the discounted interest or the bill stamp tax, the interest computation method (English, German, or French), the availment restrictions (7), the currency together with the buying rate and expected selling rate (only for k=5), and a symbol for credits bearing a combined restriction amountwise. A number of sequentially ordered triplets are also inputed, where the first two components characterize the period of time in which the ceiling given by the third component applies.

The master file is updated by the program I in Figure 2 before each run which normally takes place at the end of a decade. Updated amounts are inputed, and changes in the composition of credits available and other planning data may be made after each run.

Based on the updated master file the coefficients of the constraint matrix are generated in MPS format (Ref. 1) via program II (see Fig. 2). The MPS input

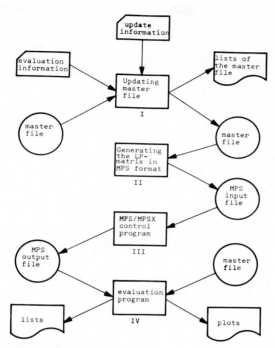

FIGURE 2. THE COMPUTER PROGRAMMING SYSTEM

serves as the input file for the programming system MPS or alternatively for the
newer version MPSX (Ref. 2). Either system solves the linear program and needs a
simple control program III written in the MPS command language that also yields
the results in standardized form as an MPS output file. An evaluation program IV
uses this MPS output file by means of READCOMM (Ref. 3), and using the master file
produces lists and plots which represent the linear programming solution in a
form which can be understood by financial managers.

    In summary, the following lists are produced:
- Lists of the updated master file.
- Lists of new availments, repayments, and the existing amounts in each loan
    category at the beginning of each decade during the planning period.
- Lists of the expected interest payments and receipts at the beginning
    of each decade.
The following plots may be provided via a printer:
- Total loan amount required and total limit of credit alternatives in the
    planning period.

- Availments and limits for each credit type.
- Availments and limits for each single credit line from various banks.
- Availments and limits for each given interest statement.

The above lists may be useful to the finance manager as a decision aid for money to be borrowed or deposited. In addition, these results can be used in a simulation model for testing new, and possibly improved, strategies. The plots also allow us to instantly recognize if, for example, there is enough (generally cheaper) advance financing available or if credits bearing favorable interest rates are available in sufficient amounts.

Finally, it should be noted that the programs I, II, and IV are written in PL/I. The programming system shown in Figure 2 is run on an IBM 370/165 computer.

In our firm we have used the computer programming system to compare the total amount of interest paid and received during one fiscal year based upon the actual composition of credits as chosen by the finance manager to that composition suggested by the program. We arrived at an interest savings of 6.6%. Even when admitting that some of the decisions made by the finance manager were influenced by overall business viewpoints and that he did not exactly know the future changes of the data at the time of his decisions, half of the percentage stated above represents a rather considerable amount. The corresponding linear program for a fiscal year had 1,170 rows corresponding to about 35 bank credits. To generate its coefficients, solving the linear program (with MPS), and evaluating its solution time required, using an IBM 370/165 computer, about 10 minutes CPU time and a region of approximately 150K. Using MPSX we needed somewhat less than half of the computing time. About 90% of the overall time was spent solving the linear program.

4. *Examples, Interpretations of the Results*

For illustrative purposes, we consider four loan compositions representing four possible optimal linear programming solutions. The values of the variables in these solutions versus time are shown in Figures 3 to 6. In each case, the planning period is one year or 36 decades. Initially there are no availments. In all four cases the same credit demand curve, which is typical for mail-order houses like our concern, is used. In particular, for each decade i (i=1,...,36) the demand is $\sum_{h=0}^{i-1} (a_h - e_h)$ (the total amount of credit at period i).

Figures 3 to 6 are explained as follows. The available credit alternatives in all cases are shown at the bottom of the figures. Each alternative is characterized by a three digit number, the first of which stands for the credit type according to (1). If the corresponding box is not filled with a pattern it means that the corresponding credit is not in the optimal solution.

Duration and actual annual interest rate are assigned to each credit. Credit limits are represented by one or more triplets (C_h, D_h, S_h) meaning that in the time interval $[C_h, D_h]$ the credit limit S_h must not be exceeded. If there are combined credit ceilings this is indicated by a bracket (}).

The existing credit availments are shown above and existing deposits are shown below the time axis, respectively. The difference of the corresponding rectangle in one decade is then as high as the rectangle of the credit demand. We endeavored to plot nearer to the time axis those credits carrying lower rates of interest. In general, the optimal solutions graphed are not unique.

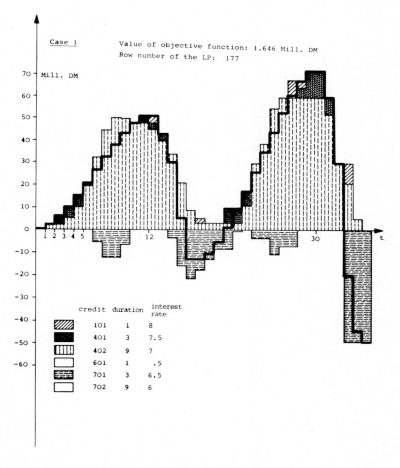

FIGURE 3.   THE OPTIMAL SOLUTION FOR ILLUSTRATION 1

In Figure 3 all the credits are available throughout the whole year.   In
essence the credit 402, with a duration of 9 decades and bearing the most favorable interest rate, is being used.   Only some peak requirements are covered by
credits 401 and 101.   The rectangles of 402 above the demand curve result from the
fact that a short term time deposit 701 with a favorable interest rate is available.

   As opposed to Figure 3 there are in Figure 4 no attractive (from an interest
viewpoint) time deposits but only a credit balance in current account 601 carrying
.5% interest that is used only in the decades 17 to 20 and 34 to 36 when the receipts predominate.   For time deposits there is an upward trend in the interest

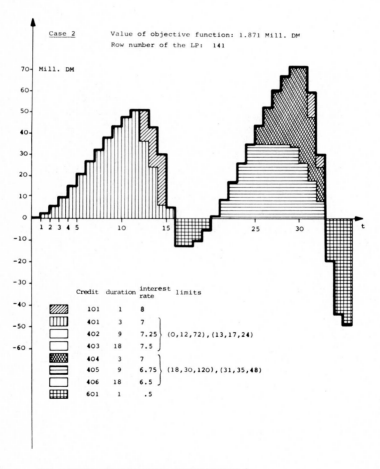

FIGURE 4.   THE OPTIMAL SOLUTION FOR ILLUSTRATION 2

rates with increasing duration during the first 6 months. This tendency is reversed in the second half of the year. In Figure 4 the overdraft 101, though less
favorable, is used during some specific decades because interest is paid for only
one decade. Compared to Figure 3 the total amount of interest is essentially
higher because the placement of time deposits is not given.

Figures 5 and 6 only differ in the amount of the credit limits for advances
401 to 406. Note the difference in the total amount of interest. In Figure 5 the
limit during the decades 7 to 11 and 24 to 29 is reached so that a relatively high
portion of requirements must be covered by the overdraft 101. As opposed to
Figure 4, Figures 5 and 6 show that the advance 406 with a duration of 18 decades

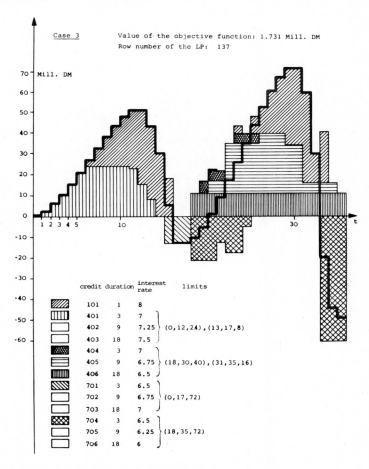

FIGURE 5. THE OPTIMAL SOLUTION FOR ILLUSTRATION 3

is used because the surplus liquidity prevailing during the last three decades can be placed on time deposits at favorable interest rates.

The four problems required approximately a total of 30 seconds of computing time using an IBM 370/165 computer with MPS.

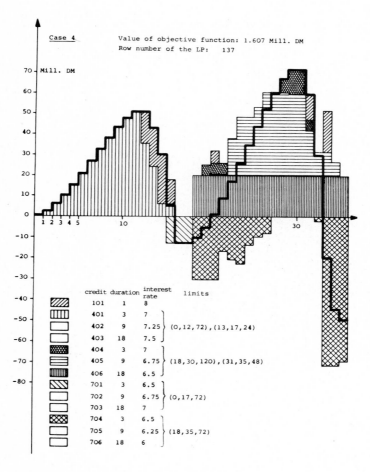

FIGURE 6.   THE OPTIMAL SOLUTION FOR ILLUSTRATION 4

References

1. MPS/360; Linear and Separable Programming - User's Manual, IBM Form H-20-0476.

2. MPSX; Linear and Separable Programming - Program Description, IBM Program Number 5734-XM 4.

3. MPS/360; Read Communications, Format (READCOMM) - Program Reference Manual, IBM Form GH-20-0372.

4. Weinberg, F., and M. Rössler; "Mathematische Optimierungsmehtoden in der Investitionsplanung," Unterlagen für einen Kurs des Instituts für Operations Research der ETH-Zurich, Mai 1970.

PART II

APPLICATIONS FROM AGRICULTURAL ENVIRONMENTS

Chapter 8

EXPANSION PLANNING FOR

A LARGE DAIRY FARM

William Swart
Department of Management Science
University of Miami
Coral Gables, Florida   33124

Calvin Smith
West Virginia College of Graduate Studies
Institute, West Virginia   25112

and

Thomas Holderby
West Virginia College of Graduate Studies
Institute, West Virginia   25112

Abstract.   The objective of this study was to
build a model of a large dairy farm operation
from which an optimal operative policy over time
could be developed and analyzed. The problem
was typical of a small business expanding into
a medium sized one in that accounting and arithmetic were no longer adequate for control and
optimization. The owner of the dairy farm hired
a private consultant to help plan expansion of
the operation.

Time staged linear programming provided an
operating and long-range growth plan which in-
tegrated a proper balance of crop growing and
purchasing with a cattle herd whose growth is
an exponential function. The results indicated
that productivity could be tripled and profit
increased ten-fold.

Currently, the model is being used at a large
dairy farm in the Ohio Valley to determine the
acreage of each crop to plant each year, the
amount of grain and hay to purchase each year, the
disposition of new-born cattle so as to maintain
an optimal livestock mix, together with projected
growth rates and profits for future years. The
farm foreman particularly liked using the linear
programming model because it assured a fair evaluation of new ideas. He finds it draws more on
employee's imagination to innovate when there is
a means to evaluate them in light of the total
operation.

1. *Introduction*

The specific dairy farm for which this study was performed is located in the
Ohio Valley. It had, at the time, a herd of 1000 head of cattle which made it
one of the larger dairy farms in the area.

During the previous year, the owner of the farm encountered problems that he
could not solve, to his satisfaction, through the use of traditional methods,
based on accounting and arithmetic, which had proven satisfactory in the past.
As the owner saw it, his problems included:

1. A three year surplus of some farm grown feeds, while other feeds
    were being purchased that could be grown with additional capital
    investment.

2. Decisions on building new barns, and other items requiring capital
    expenditures, were being delayed because of uncertainties concerning
    the rate and direction of growth and capacity.

3. Several alternatives were available, such as whether to rent land,

keep bull calves for beef, and whether to build storage facilities.

To select an appropriate course of action in order to remedy the above and
other problems, the owner could exercise control over two basic but interacting
resources, namely the existing lands, used exclusively for growing various feeds
for the cattle, and the cattle. The size of the herd of cattle changes over time
due to births and attrition rates which are an integral and unavoidable factor in
the operation of a dairy farm.

For feeding purposes, cattle are divided into four groups. Group 1 cattle are
the youngest (0-3 months), while group 4 cattle are fully grown and milk producing cows. Each group is fed varying proportions of grain corn, silage, hay, and
haylage. Of these crops, the only ones which the owner wished to consider for
purchase were grain corn and hay. Of the other crops, enough must be produced
to meet the feeding requirement of the entire herd, although it was possible to
store certain quantities of all crops, except grain corn, from year to year.

In order to control the size of the herd, it was possible to sell a portion
of the young heifers and bulls at various points in their development. Typically, new born calves could be sold as well as some group 3 heifers, while all
bulls were disposed of before or during the time they were of group 3 age. The
sale of full grown cows was not considered to be an alternative since there is
virtually an unlimited demand for milk and thus yearly milk revenues make their
sale unprofitable.

The owner of the dairy farm wished to have his questions answered so as to be
able to expand the current livestock herd while making best use of the crop resources and utilizing a minimum amount of new capital. It was the objective of
this study to construct a model of the combined dairy and farm operations that
would yield specific numbers of cattle to be kept, the optimum time for selling,
a policy for determining what crops are to be planted on available acres, and
guidelines for grow or buy decisions. In addition, a model integrating a growing livestock herd with use of croplands would allow farm management to investigate alternatives dealing with such choices as the utilization of barn space and
the rental of additional lands.

2. The Model, Formulation

In order to develop a model, it was helpful to visualize the dairy farm as
shown in Figure 2. The progression of cattle through the first two groups requires one year while the progression through group 3 into group 4, the milk
producing group, takes one more year. At each time stage the farm must grow
and purchase enough crops to meet the feeding requirements of the entire herd.
We now develop a set of relationships describing the cattle sector; the crop

sector is then modelled, and includes the interaction with the cattle sector.

The Cattle Subsystem

When focusing upon the progression of cattle through time, the relationship of the number of milk-producing cows (group 4 cattle) to cattle of other groups is dependent on the number and type of cattle kept and sold during each period. In particular, the relationships can be represented as in Figure 1,

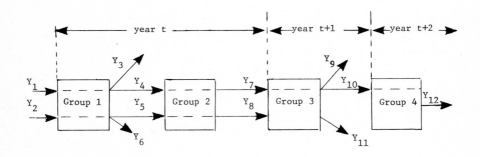

FIGURE 1: GRAPHICAL REPRESENTION OF THE CATTLE SUBSYSTEM

where the variables are defined below.

Y_1 —   The number of group 1 heifers born.

Y_2 —   The number of group 1 bulls born.

Y_3 —   The number of group 1 heifers sold at birth.

Y_4 —   The number of group 1 heifers not sold at birth.

Y_5 —   The number of group 1 bulls not sold at birth.

Y_6 —   The number of group 1 bulls sold at birth.

Y_7 —   The number of group 2 heifers.

Y_8 —   The number of group 2 bulls.

Y_9 —   The number of group 3 heifers sold.

Y_{10} —   The number of group 3 heifers not sold.

Y_{11} —   The number of group 3 bulls.
          (Since the owner does not wish to keep any mature bulls, all group 3 bulls are sold.)

Y_{12} —   The size of group 4 cattle (i.e., the size of the milk-producing) herd.

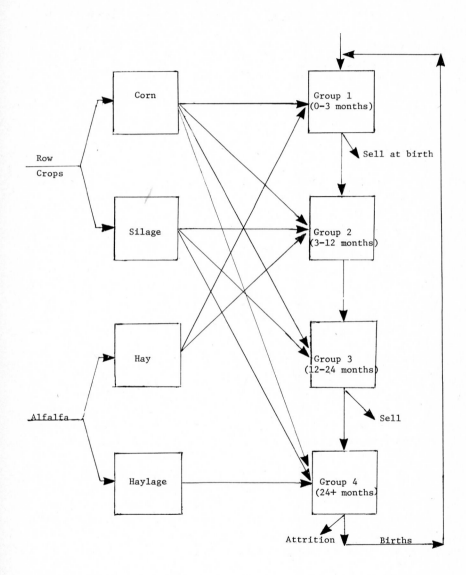

FIGURE 2:   CONCEPTUAL REPRESENTATION OF THE DAIRY FARM

The relationships between these variables, as conceptualized by Figure 1, are dependent on the particular time period, t, under consideration. During each time period (year), each milk producing cow (group 4 cattle) has one calf, and approximately half of all calves born will be bulls, and the other half heifers. Consequently, the following relations will hold at each period t (1 period = 1 year):

$$Y_{3,t} + Y_{4,t} = 0.5\ Y_{12,t} \qquad (1)$$

$$Y_{5,t} + Y_{6,t} = 0.5\ Y_{12,t}. \qquad (2)$$

Calves are not sold while they are of group 2 age. Also, it should be recalled that the progression from group 1 to group 2 is made in the same year. Hence, for each t, we have the following equations:

$$Y_{7,t} = Y_{4,t} \qquad (3)$$

$$Y_{8,t} = Y_{5,t}. \qquad (4)$$

Group 2 cattle will become group 3 cattle in the next period. Since all bulls of that age must be sold, for each t, we have the equalities:

$$Y_{10,t+1} + Y_{9,t+1} = Y_{7,t} \qquad (5)$$

$$Y_{11,t+1} = Y_{8,t}. \qquad (6)$$

Finally, group 4 cattle suffer approximately a 30% attrition rate each year, but at the same time the group 4 population is enlarged by the infusion of the previous period group 3 heifers that were kept. Consequently,

$$Y_{12,t+1} = Y_{10,t} + 0.7\ Y_{12,t}. \qquad (7)$$

Equations (1) through (7) embody all the relationships necessary to describe the cattle subsystem of the farm with the exception of some initial conditions indicating the number of cattle currently existing within each group. We now describe the crop subsystem together with the crop-cattle interactions.

The Crop Subsystem

A large portion of the crop subsystem can be adequately represented by a series of equations, each of which equates the amount of a certain crop grown

during a particular period plus the amount available in storage to the amount
that will be consumed in that period plus the amount that is placed in storage
for use in subsequent periods. Schematically, this material balance is exhibited
in Figure 3.

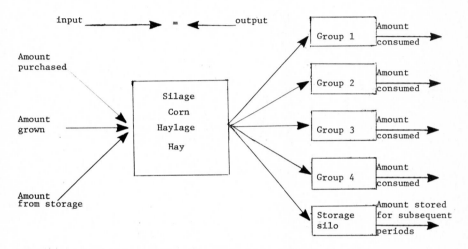

FIGURE 3. SCHEMATIC OF MATERIAL BALANCE EQUATIONS FOR THE CROP SUBSYSTEM

In addition to the material balance equations, it is also necessary to incorporate the relations describing the limitations that exist on the amount of each
crop that can be stored due to silo or barn capacity.

Let the variables under control of the farm owner be defined as below.

$X_{1,t}$ — The number of acres devoted to growing silage in year t.

$X_{2,t}$ — The number of acres devoted to growing corn in year t.

$X_{3,t}$ — The number of acres devoted to growing haylage in year t.

$X_{4,t}$ — The number of acres devoted to growing hay in year t.

$X_{5,t}$ — The number of bushels of corn purchased in year t.

$X_{6,t}$ — The number of bales of hay purchased in year t.

$z_{1,t}$ — The number of tons of silage in storage at the end of year t.

$z_{2,t}$ — The number of tons of haylage in storage at the end of year t.

$z_{3,t}$ — The number of bales of hay in storage at the end of year t.

$W_{i,t}$ — The total consumption of crop i in year t by all groups of cattle.

The $W_{i,t}$ (i=1,2,3,4) are variables which relate the cattle subsystem variables to the crop subsystem variables.

Then, the equation describing the material balance for silage is given
by

$$z_{1,t} + 20\, X_{1,t} - W_{1,t} - z_{1,t+1} = 0, \tag{8}$$

where

$$W_{1,t} = 2.45(Y_{7,t} + Y_{8,t}) + 4.9(Y_{9,t} + Y_{10,t} + Y_{11,t}) + 5.6\, Y_{12,t} \tag{9}$$

The coefficient of $X_{1,t}$, 20, indicates that each acre yields 20 tons of silage,
the quantity $W_{1,t}$ represents the yearly consumption of silage by the various cattle groups and the coefficients in (9) represent the per cattle consumption of
silage in tons for each group of the herd. (Group 1 cattle do not consume silage.)

The storage limitation for silage is equal to the capacity of the storage
silo which is 18,000 tons. Thus,

$$z_{1,t} \leq 18,000. \tag{10}$$

Similarly, we can define the restrictions for the other crops as shown below.
For corn:

$$X_{5,t} + 150\, X_{2,t} - W_{2,t} = 0 \tag{11}$$

$$W_{2,t} + 4(Y_{4,t} + Y_{5,t}) + 17(Y_{7,t} + Y_{8,t}) + 4.8(Y_{9,t} + Y_{10,t} + Y_{11,t}) + 93.2\, Y_{12,t} \tag{12}$$

$$150\, X_{2,t} \leq 38,600. \tag{13}$$

For haylage:

$$z_{2,t} + 12\, X_{3,t} - W_{3,t} - z_{2,t+1} = 0 \tag{14}$$

$$W_{3,t} = Y_{12,t} \tag{15}$$

$$z_{2,t} \leq 2,400. \tag{16}$$

For hay:

$$z_{3,t} + 100\ X_{4,t} + X_{6,t} - W_{4,t} - z_{3,t+1} = 0 \qquad (17)$$

$$W_{4,t} = 5(Y_{4,t} + Y_{5,t}) + 14\ Y_{8,t} \qquad (18)$$

$$z_{3,t} \leq 12{,}000. \qquad (19)$$

In addition to the previous relations, certain operating conditions must be satisfied. In particular, it was necessary for the farm to continously grow crops on all 530 acres it had available for cultivation since any land not used for growing crops during a given year became unsuitable for growing crops for a certain time thereafter. As an alternative for expanding the growing capacity, the owner could decide to rent up to 160 acres at any point in time. Hence his available capacity, including the restriction of continous use for 530 acres, can be expressed by

$$X_{1,t} + X_{2,t} + X_{3,t} + X_{4,t} + V_t = 690 \qquad (20)$$

$$V_t \leq 160 \qquad (21)$$

where V_t represents the number of acres _not_ rented in period t. (Note that when all 160 acres are rented, $V_t = 0$, while if none are rented $V_t = 160$.) Finally, as an operating condition, the owner wished to devote at least 80 acres to the growing of alfalfa (both hay and haylage are alfalfa crops). This condition gives

$$X_{3,t} + X_{4,t} \geq 80. \qquad (22)$$

The Measure of Effectiveness

A specific assignment of values to the variables defined in the cattle and crop subsystems which satisfy relations (1) through (22) as well as the necessary initial conditions and nonnegativity requirements constitute a feasible operating policy for the dairy farm. Since it is very likely that more than one set of values for the decision variables exists that satisfies all relations derived so far, it is necessary to develop a criterion, or measure of effectiveness, whereby the relative "goodness" of one feasible operating policy with respect to others can be established.

After careful deliberation, the owner of the dairy farm stated that an acceptable criterion for selecting a feasible operating policy for the farm would be

the resulting profit, which he wished to maximize. For the purposes of the study,
it was appropriate to define profit as the difference between the sum of revenues
and the sum of all the costs associated with the decision variables. The amount
of revenue or cost associated with each decision variable is exhibited in Table 1.
These values were obtained from the General Ledger, which listed all farm expenses
and from drawing upon the experience and practical knowledge of the owner and
employees. The costs assigned to storing crops were derived by considering the
cost of growing an acre of that crop, dividing the cost by the tonnage or bales
yielded per acre, and multiplying this cost by the interest rate that would have
been applied if this money had been deposited in a bank. Therefore, this cost
simply represents the investment return foregone by using the money to cultivate
land as opposed to depositing it in a bank. The profit for year t, R_t, is then
given by the sum of the products of the variables times their corresponding appropriate cost or revenue; that is,

$$
\begin{aligned}
R_t = {} & 35\ (Y_{3,t} + Y_{6,t}) + 500\ Y_{9,t} + 258.3\ Y_{11,t} + 662\ Y_{12,t} \\
& - 54.9\ (Y_{4,t} + Y_{5,t}) - 23.5\ (Y_{7,t} + Y_{8,t}) - 18.7\ Y_{10,t} \quad (23) \\
& - 87.2\ X_{1,t} - 97.2\ X_{2,t} - 67.4\ X_{3,t} - 68.4\ X_{4,t} - 1.5\ X_{5,t} \\
& - 2.25\ X_{6,t} - 0.39\ Z_{2,t} - 0.30\ Z_{1,t} - 0.047\ Z_{3,t}.
\end{aligned}
$$

The Model

Relations (1) through (23) comprise a representation of the dairy farm for a
particular period t. However, these relations have to be satisfied for every year
the dairy farm is in operation. Also, the criteria by which the owner wished to
operate the dairy farm was not the maximization of profits during a single year t,
but the maximization of total profits over the life of the farm. For convenience
and in order to keep the problem from becoming too large, it was decided to represent the dairy farm for a period of 25 years, and to determine the optimal operating policies for each of these 25 years. For a single period, the coeficient
matrix is exhibited in Figure 4. (Note that the expressions for $W_{i,t}$ have been
substituted in the material balance relations.) The complete 25 year linear
programming model of the dairy farm would consist of 25 blocks of data, similar
to that shown in Figure 4, with the overall objective of maximizing the total profit, where profits in the future are adjusted by an interest factor to reflect the
time value of money. In short, the objective is to maximize the present worth of
the farm profit.

	PERIOD t																				PERIOD t + 1							
	Y_3	Y_4	Y_5	Y_6	Y_7	Y_8	Y_9	Y_{10}	Y_{11}	Y_{12}	X_1	X_2	X_3	X_4	X_5	X_6	z_1	z_2	z_3	V	R	Y_9	Y_{10}	Y_{11}	Y_{12}	z_1	z_2	z_3	
CATTLE SECTOR	1	1								-0.5																			= 0
			1	1						-0.5																			= 0
		1			-1																								= 0
			1			-1																							= 0
					1																	-1	-1						= 0
						1																		-1					= 0
								1		0.7															-1				= 0
CROP SECTOR					-2.45	-2.45	-4.9	-4.9	-4.9	-5.6	20						1									-1			= 0
		-4	-4		-17	-17	-4.8	-4.8	-4.9	-93.2		150			1														= 0
										-2.2			12					1									-1		= 0
		-5	-5		-14	-14								100		1			1									-1	= 0
																	1												≤ 18,000
																		1											≤ 2,400
																			1										≤ 12,000
												150																	≤ 38,500
FARM OPERATING											1	1	1	1						1									= 690
													1	1						1									≤ 160
																													≥ 80
PERIOD RETURN FUNCTION	35	-549	-549	35	-235	-235	500	-187	2583	662	-872	-972	-674	-684	-15	-225	-031	-039	-005	0	-1								

FIGURE 4. TYPICAL SINGLE TIME PERIOD FORMULATION

Variables	Description of Revenue or Cost	Amount of Revenue or cost ($)
$Y_{3,t}$, $Y_{6,t}$	Group 1 cattle can be sold at market price	+35
$Y_{9,t}$	A group 3 heifer can be sold at market price	+500
$Y_{11,t}$	A group 3 bull can be sold for beef	+258.3
$Y_{12,t}$	A cow yields revenue from her milk	+662
$Y_{4,t}$, $Y_{5,t}$	The care and other expenses of group 1 cattle	−54.9
$Y_{7,t}$, $Y_{8,t}$	The care and other expenses of group 2 cattle	−23.5
$Y_{10,t}$	The care and other expense of a group 3 heifer	−18.7
$X_{1,t}$	Cost of growing 1 acre of silage	−87.2
$X_{2,t}$	Cost of growing 1 acre of grain corn	−97.2
$X_{3,t}$	Cost of growing 1 acre of naylage	−67.4
$X_{4,t}$	Cost of growing 1 acre of hay	−68.4
$X_{5,t}$	Cost of purchasing 1 bushel of grain corn	− 1.5
$X_{6,t}$	Cost of purchasing 1 bale of hay	− 2.25
$Z_{1,t}$	Cost of storing 1 ton of silage 1 year	−(0.07)(67.4)/12
$Z_{2,t}$	Cost of storing 1 ton of haylage 1 year	−(0.07)(87.2)/20
$Z_{3,t}$	Cost of storing 1 bale of hay 1 year	−(0.07)(68.4)/100

TABLE 1. REVENUE AND COST INFORMATION FOR THE MODEL

3. *Computational Results*

The linear programming model developed in the previous section is specialized
in the sense that basically the same process recurs from one period to another,
but in which each period's decision problem is dependent on the decisions of the
previous period. This type of linear programming problem is normally referred to
as a time staged linear program or as a dynamic linear programming problem. Since
this type of problem arises frequently, it is discussed in several books (e.g.,
see Ref.2).

The nature and source of the data used for our problem has been discussed,
along with the model development, in the previous section. To solve the dairy
problem, a commercially available linear programming computer system developed by
IBM (Ref.1) was used. The results obtained are tabulated in Table 1 and are
graphically displayed in Figure 5.

A preliminary analysis of the results indicated that, as a profit maximizing
strategy, the number of cattle in group 4 should be allowed to increase as rapidly as possible. Since initially more crops can be grown than consumed, quanti-

ties in storage increase while the available acreage is devoted to growing corn
and hay (items that could be purchased) as well as haylage and silage. During
year 7, the acreage that was devoted to corn and hay is shifted to haylage and
silage, and the additional planting increases the rate at which silage is stored
since, at this time, there is no additional capacity to store haylage. By year 11,
it becomes necessary to devote to haylage such a large portion of the acreage al-
lotted to silage that some of the feeding requirements are satisfied from previously stocked silage. Finally, by year 15, the group 4 cattle become so numerous
that the accumulated storage is depleted and it becomes necessary to reduce the
number of cattle in this group. To accomplish this in an optimal manner, group 1
heifers should begin to be sold in period 14, while it will also be necessary to
sell a few heifers of group 3 in year 15. After this initial herd depletion,
which is directly a consequence of the hoarding of crops during the first 11 years,
the group 4 herd size becomes more stable and is regulated by only keeping enough
group 1 heifers to maintain the group 4 cattle population, while all other calves
are sold at birth.

According to the results from the solution to the linear programming problem,
the above described strategy for running the farm is "optimal". When presented
to the owner, though, the reaction was one of incredibility and displeasure fos-
tered primarily by the requirement that the owner should reduce his herd size at
various points in time. In short, as often happens when formulating linear programming models, the initial version of the model did not capture all of the considerations which the owner considered important, but thought to be self evident
to anybody (in the dairy business, that is). After consultations and discussions
it was felt with relative certainty that the primary modification required for the
model to satisfy the dairy farm owner would be a statement to the effect that the
herd size should not decrease at any point in time. This requirement, although
not desirable from a strict economic standpoint, was very desirable for the image
and reputation of the dairy farm amongst its peers in the industry, where stability and steady growth were synonymous with prosperity and good management.

To incorporate the requirement of a non-decreasing group 4 herd size, the
constraints

$$Y_{12,t} \leq Y_{12,t+1} \quad \text{for all } t \tag{24}$$

were added to the model. The solution obtained for this revised version of the
dairy farm model is tablulated in Table 2 and shown graphically in Figure 6. As
can be seen, the solution to this model exhibited the stability desired by the
owner, and he found this solution to be much more acceptable than the one obtained
from the original model.

t	Cattle Subsystem					z_1	z_2	z_3	Crop Subsystem						R
	Y_3	Y_4,Y_7	Y_9	Y_{10}	Y_{12}				X_1	X_2	X_3	X_4	X_5	X_6	
1	0	159	0	154	317	4200	1000	1500	0	224	175	131	0	0	155886
2	0	188	0	159	376	1326	2400	11602	164	257	69	40	1142	0	187514
3	0	211	0	188	422	1318	2400	12000	315	257	77	40	6025	0	195140
4	0	242	0	211	483	3885	2400	12000	298	257	89	46	12508	0	223892
5	0	274	0	242	549	5586	2400	12000	280	257	101	52	19489	0	254880
6	0	313	0	274	626	6330	2400	12000	289	257	115	29	27616	0	290911
7	0	356	0	313	713	6582	2400	8946	559	0	131	0	75396	5535	276503
8	0	406	0	356	812	11472	2400	2176	541	0	149	0	85885	8783	310192
9	0	462	0	406	924	15124	2400	0	521	0	169	0	97817	10003	355679
10	0	526	0	462	1053	17365	2400	0	497	0	193	0	111415	11394	413570
11	0	600	0	526	1199	18000	2400	0	470	0	220	0	126899	12977	480086
12	0	683	0	600	1366	16805	2400	0	440	0	250	0	144537	14781	556448
13	0	778	0	683	1556	13526	2400	0	405	0	285	0	164625	847	644016
14	842	45	0	778	1772	7873	2400	0	365	0	325	0	169834	0	902236
15	1009	0	45	0	2018	1578	443	0	483	0	207	0	188333		1089607
16	36	671	0	0	1413	0	0	0	468	0	222	0	145765	12740	605864
17	234	260	0	671	989	0	510	0	466	0	224	0	100860	4942	429432
18	299	383	0	260	1363	0	0	0	483	0	207	0	136303	7272	624890
19	263	344	0	383	1214	0	0	0	467	0	223	0	122225	6543	546507
20	264	352	0	344	1232	0	0	0	464	0	226	0	123929	6691	556430
21	172	431	0	352	1207	0	0	0	469	0	221	0	123260	8194	527140
22	288	310	0	431	1197	0	0	0	471	0	219	0	120167	5897	542146
23	351	284	0	310	1269	0	0	0	457	0	233	0	125743	5388	590742
24	0	599	0	284	1199	0	0	0	470	0	220	0	125685	11389	492696
25	310	251	171	428	1123	0	0	0	484	0	206	0	112797	4771	595945

TABLE 1.   COMPUTATIONAL RESULTS

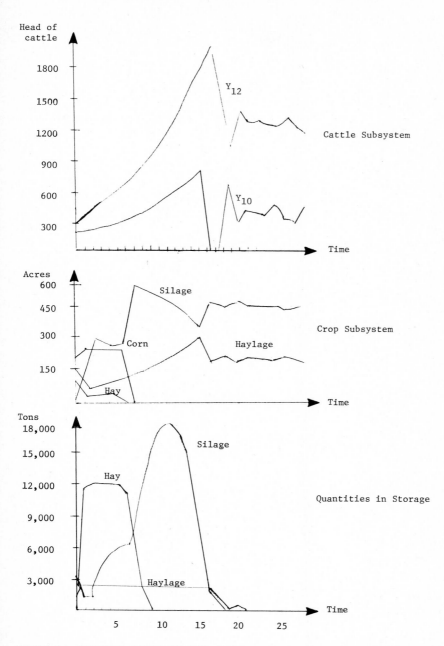

FIGURE 5.   GRAPHICAL REPRESENTATION OF THE SOLUTION TO THE ORIGINAL MODEL

t	Cattle Subsystem					Crop Subsystem									R
	Y_3	Y_4, Y_7	Y_9	Y_{10}	Y_{12}	Z_1	Z_2	Z_3	X_1	X_2	X_3	X_4	X_5	X_6	
1	0	159	0	154	317	4200	1000	1500	0	224	175	131	0	0	155886
2	0	188	0	159	375	1326	2400	11602	164	257	69	40	1142	0	187514
3	0	211	0	188	422	1318	2400	12000	315	257	77	40	6025	0	195140
4	0	242	0	211	483	3885	2400	12000	298	257	89	46	12058	0	223892
5	0	274	0	242	549	5586	2400	12000	280	257	100	52	19489	0	254880
6	0	313	0	274	626	6330	2400	12000	289	257	115	29	27616	0	290911
7	0	356	0	313	712	6582	2400	8946	559	0	131	0	75396	0	270503
8	0	406	0	356	812	11472	2400	2176	541	0	149	0	85885	5535	310192
9	0	462	0	406	924	15124	2400	0	521	0	169	0	97817	8783	355679
10	0	526	0	462	1053	17365	2400	0	497	0	193	0	111415	10003	413570
11	0	600	95	462	1199	18000	2400	0	470	0	220	0	126899	11394	529103
12	0	636	218	432	1272	16805	2400	0	457	0	233	0	134737	12080	625477
13	0	636	254	381	1272	14504	2400	0	457	0	233	0	134910	12080	653979
14	0	636	254	381	1272	12026	2400	0	457	0	233	0	134910	12080	644748
15	0	636	254	381	1272	9548	2400	0	457	0	233	0	134910	12080	645515
16	0	636	254	381	1272	7071	2400	0	457	0	233	0	134910	12080	646284
17	0	636	254	381	1272	4593	2400	0	457	0	233	0	134910	12080	647052
18	254	381	254	381	1272	2115	2400	0	457	0	233	0	129570	7248	695533
19	254	381	254	381	1272	260	2400	0	474	0	216	0	128349	7248	570441
20	254	381	0	381	1272	0	2191	0	487	0	203	0	128349	7248	570346
21	254	381	0	381	1272	0	1826	0	487	0	203	0	128349	7248	570481
22	254	381	0	381	1272	0	1461	0	487	0	203	0	128349	7248	570631
23	254	381	0	381	1272	0	1095	0	487	0	203	0	128349	7248	570774
24	254	381	0	381	1272	0	730	0	487	0	203	0	128349	7248	570916
25	254	381	0	381	1272	0	365	0	487	0	203	0	128349	7248	571058

TABLE 2.   SOLUTION FOR THE REVISED MODEL

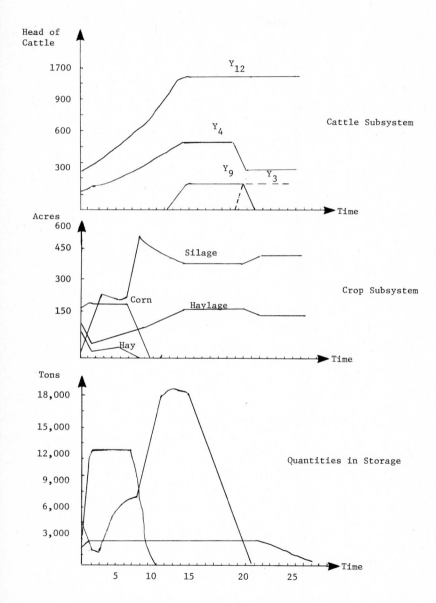

FIGURE 6. GRAPHICAL REPRESENTATION OF THE SOLUTION TO THE REVISED MODEL

4. Conclusions

The results from the computer run of the revised model indicated a somewhat different strategy than the one obtained from the initial model. The most significant difference was the time and manner in which herd sizes were to be checked. The revised model indicated that a policy of maximum growth should be adopted for the first 11 years of the planning horizon. This growth rate was to be stopped by initiating the sale of group 3 heifers in year 11. Enough group 3 heifers were to be sold each year so as to maintain the herd size of group 4 cattle at 1272 head. By year 18 the quantity of crops in storage is forecasted to become depleted, which necessitates making all feeds available to milk producing cows, hence no more heifers are sold at the group 3 stage and, instead, those heifers which will not be eventually added to the milk producing herd are sold at birth. (This steady state strategy was advocated directly by the initial model.) From a crop subsystem standpoint, the last year in which acreage is devoted to growing hay is year 6, while the last year corn is grown is in year 7. Thereafter, these are purchased on the market. Enough haylage is grown on and after year 7 to feed the cattle and to keep the silos full, while the rest of the acreage is used to grow silage. In year 19, when the supply of silage is dwindling, more land is devoted to growing silage.

This strategy, and in particular the stability it implied with respect to herd size and acreage devoted to crop growing, did ease the owner's concern regarding the fluctuations in herd size and variability in crop growing policies. Nevertheless, the owner, at first, could not accept the model's apparent optimism in arriving at an unexpectedly favorable capacity and profit position. He gave a a few clouded reasons as to why he could not agree, but promised to think it over.

A short time later, he reported that he was ready to reconsider the findings because he had provided the input and, therefore, had been rejecting his own data. After this ragged beginning, the results were carefully reviewed. In particular, the stream of returns associated with the policies derived from the initial and revised models were compared, as shown graphically in Figure 7, in order to evaluate the potential money loss associated with the revised model. From these results, it was decided to use the revised model for policy making decisions for the dairy farm.

The first positive action as a result of the study involved two long-delayed decisions on capital expenditures. The capacity of a new calf barn was determined and the decision to provide additional corn storage was made and construction began immediately.

With an overall plan based on the linear programming model, it became possible

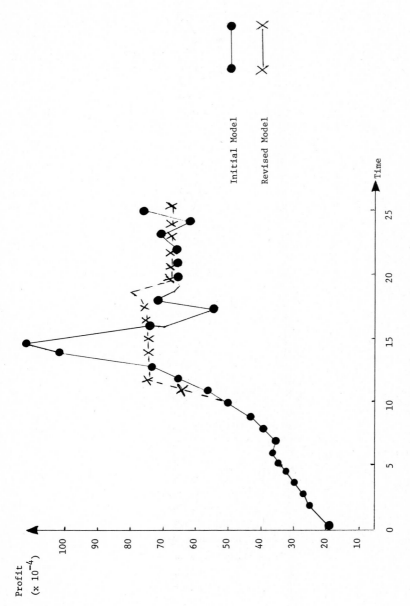

FIGURE 7.   COMPARISON OF YEARLY PROFITS, ORIGINAL AND REVISED MODELS

to forecast cash flow, capital requirements, profit, and manpower requirements.

Prior to the use of linear programming, there had been no comprehensive approach to evaluation of alternative investment opportunities. Now the farm is quantitatively balancing buy or sell decisions on cattle and feedstocks. Since the farm is adjacent to an urban area, there are each year increasingly favorable opportunities to convert land to other uses. The model will be beneficial in helping select the proper direction and time for changes in land use.

The farm foreman particularly likes using the linear programming model because it assures a fair evaluation of new ideas. He finds it draws more on employees' imagination to innovate when there is a means to evaluate them in light of the total operation.

The by-products of the study were highly usable by the farmer. The rigorous requirement to classify, calculate costs and yields, and develop connecting relationships caused the farm management to see the business in new ways. From this came new insight and perception that brought improvements in operating procedures, and assigned responsibility and personnel policy.

Perhaps the greatest benefit of this study is the confidence that farm management feels now that they know where the business is headed and how to make the best decision in an environment which is becoming increasingly complex. This confidence is apparent to farm employees, its bankers, and the business public with whom it deals.

References

1. Introduction to Mathematical Programming System, IBM Corporation Newsletter No. Gm20-2452, IBM, White Plains, N.Y. 10604, September 6, 1971.

2. Wagner, H.; Principles of Operations Research with Application to Managerial Decisions, Prentice-Hall,Englewood Cliffs, New Jersey, 1969.

Chapter 9

LINEAR PROGRAMMING ANALYSIS OF A CATTLE FEEDLOT*

Said Ashour
The ADAR Corporation
233 East Lancaster Avenue
Ardmore, Pennsylvania   19003

and

Carl Anderson
Department of Agricultural Engineering
Iowa State University
Ames, Iowa   50010

*This research was conducted while the authors were at Kansas State University,
Manhattan, Kansas.

Abstract.   The problem of handling animal wastes
has had a major influence on the operation of beef
cattle feedlots in the High Plains. The best long
range method appears to be waste recycling where
the waste is returned to the land that has produced
the feed used in the feedlots. This chapter is
concerned with the development of a linear programming model which determines the economic limits to
feedlot development assuming that land disposal of
wastes are required by outside social constraints.
Constraints are included to insure that feeds produced constitute a nutritionally balanced ration
and to insure that all land employed for feed production is also used for waste disposal. It is
assumed that the feedlot has control over enough
land to serve its needs for waste disposal. The
model has been applied to data typical of the High
Plains area of Kansas and the solution shows more
concentrated feedlot development than currently
exist.

1. Introduction

The beef production industry has shifted dramatically in recent years from a
centralized industry in which live cattle were shipped to slaughter houses near
big markets to one in which cattle are fattened and slaughtered in the grain producing areas, with processed meat being shipped to market in refrigerated vehicles.
Modern developments in irrigation have opened up much of the High Plains area including Kansas, Oklahoma, and Texas, to the production of corn and grain sorghum
which can be used in the beef cattle feeding industry. Large feedlots have developed in this area in which as many as 100,000 cattle are kept in pens and fattened for market.

The basic objective of the feedlot industry is to produce large quantities of
low cost, high quality beef. The inputs to the industry include a supply of
young feeder cattle, feed grains, roughage, and water along with land, labor, and
capital. The process involves feeding a concentrated ration high in food energy
and protein to animals confined to pens so that they will gain body weight at a
maximum rate. The concetration of cattle in a small area combined with the unnatural diet tend to encourage animal health problems so that the operation becomes more efficient when a feedlot is large enough to afford specialists in
nutrition and animal diseases. Each animal requires about 5.5 tons of feed
per year so that feedlots gain a major advantage over other methods of producing
beef by locating the cattle near the source of feed, thus reducing transportation
costs. Also a large feedlot with modern equipment can handle about 1000 cattle
per employee, thus reducing labor cost.

In this chapter we explore the economics of land disposal of feedlot waste by
a recycling process in which feed is taken from the land, fed to the animals producing waste, which is returned to the land where it undergoes chemical and bio-
logical degradation with some of the plant nutrients in the waste serving to produce more feed. The optimum level of feedlot activity within a given area has
been sought when waste recycling was a part of the management system. No attempt
has been made to include components in the model to force recycling by economic
penalties or rewards to the farmer. It is hoped the resulting model can be used
to evaluate the optimal level of feedlot development in an area when recycling is
required by some social rather than economic constraint.

2. *The Problem*

Since the cattle feeding industry has only begun using mass-production techniques in the past ten years, many new problems have developed which have not been
adequately anticipated and planned for. The most serious of these has been the
handling of large quantities of animal wastes.

When cattle roam free on rangeland, natural body wastes are deposited randomly
over the area and generally cause no further problem. However, when cattle are
confined to a small pen area the wastes accumulate and must be handled as part of
the feedlot operation. Recent research indicates that an average weight animal in
a feedlot will produce about 60 pounds of wet waste per day (Ref. 6,8).

Historically, animal waste has been used as a fertilizer and soil conditioner
on farmland so that the natural choice of waste disposal methods is land disposal.
Because of the unnatural diet of fed beef, however, the waste may be high in mineral salts which tend to contaminate the soil and reduce its ability to support
plant growth. The wastes also may contain more nitrogen compounds than the soil
can handle and thus eventually cause a nitrogen pollution problem in runoff water
and ground water. Therefore, the application of animal wastes to the land in extreme amounts cannot be practiced without causing other economic and social problems. The animal wastes must be spread as thin as possible and still might reduce
rather than increase crop yields from the land on which it is spread. All of the
factors affecting animal waste disposal are not known, but land disposal still
appears to be most desirable from the ecological standpoint.

Current interest in protecting the quality of the environment has had a serious
effect on the feedlot industry, causing some older feedlots to relocate or go out
of business. Feedlot operators are forced now to consider this problem in their
system, and are looking for answers. The question becomes one of determining the
best method of handling the animal wastes and the optimum size operation in a given
area when waste disposal is included.

where

    c   is the annual profit per head after deducting the cost of maintaining the feedlot equipment. (The cost of feed and waste disposal is not included.)

    c_{ijk} is the production cost of a pound of feed i from source j on land used for waste disposal at rate k.

    X_{i5} is the amount of feed item i from zone 5 (an outside source).

    c_{i5} is the purchase cost of a pound of feed i from zone 5 (an outside source).

    c_{jk} is the cost of disposing one **ton of animal waste at rate k in zone j**.

Note that the first summation over i excludes both cottonseed meal (i=1) and salt (i=2), since these two feeds are purchased from outside sources. However, the second summation over i excludes corn silage (i=7) and sorghum (i=8), because all amounts required from these two feeds are grown in the available land.

The first set of constraints **requires that the feeds used constitute a nutrion-** ally balanced ration and will cause the animal to gain body weight at a maximum rate. One constraint is required for each nutrient element considered. These constraints are:

$$\sum_{i=1}^{8} \sum_{j=1}^{4} \sum_{k=1}^{2} a_{i\ell} X_{ijk} + \sum_{i=1}^{6} a_{i\ell} X_{i5} - 365 b_\ell Z \geq 0 \qquad (\ell = 1,2,\ldots,8), \qquad (2)$$

where $a_{i\ell}$ is the amount of nutrient ℓ in one pound of feed i, and b_ℓ is the minimum daily requirement of nutrient ℓ in the diet of one animal. Eight different nutritional requirements are considered in the present model: calcium, protein, total digestible nutrients, phosphorous, total dry matter, net food energy, carotene, and salt, corresponding to $\ell = 1,2,\ldots,8$, respectively.

The second set of constraints is on the content of the ration. These include the somewhat qualitative requirements of palitability and limits on the total volume of material which an animal can consume in the quest for sufficient quantities of the various nutrients. For example, a cow can probably get enough of each nutrient to satisfy minimum daily requirements by eating nothing but hay if it has a large enough stomach. However, the limit on stomach capacity demands that more expensive but more concentrated feeds be included to meet the requirements. Four constraints are included.

The silage content constraint is:

$$\sum_{i=7}^{8} \sum_{j=1}^{4} \sum_{k=1}^{2} X_{ijk} - p \left(\sum_{i=3}^{8} \sum_{j=1}^{4} \sum_{k=1}^{2} X_{ijk} + \sum_{i=1}^{6} X_{i5} \right) \geq 0. \qquad (3)$$

This constraint requires that the silage content of the ration (i=7,8) be at least the fraction p, and it ensures a ration that is attractive to the animals.

The protein supplement content constraint is:

$$x_{15} - q\left(\sum_{i=3}^{8}\sum_{j=1}^{4}\sum_{k=1}^{2} x_{ijk} + \sum_{i=1}^{9} x_{15}\right) = 0.$$ (4)

This constraint sets the protein supplement of the ratio $(i=1)$ at some portion represented by q.

The hay content constraint is:

$$\sum_{j=1}^{4}\sum_{k=1}^{2} x_{3jk} - s\left(\sum_{i=3}^{8}\sum_{j=1}^{4}\sum_{k=1}^{2} x_{ijk} + \sum_{i=1}^{9} x_{15}\right) \leq 0.$$ (5)

This constraint sets the hay content of the ratio $(i=3)$ at most the fraction s.

Constraints (4) and (5) are included to help control the source of nutrients in the ration and limit the consumption of roughage which does not contain much protein and food energy.

The total feed consumption constraint is:

$$365y z - \left(\sum_{i=3}^{8}\sum_{j=1}^{4}\sum_{k=1}^{2} x_{ijk} + \sum_{i=1}^{9} x_{15}\right) \geq 0.$$ (6)

This requirement limits the total daily feed consumption of the animal to y pounds. This number depends on the type, size, and age of the cattle.

The third set of constraints balances the use of land area for feed production and waste disposal. One of the basic assumptions of this model is that all land used for feed production in the area must also be used for disposal of animal wastes from the feedlot. These constraints are of the form:

$$\sum_{i=1}^{8} \frac{1}{\beta_{1k}} x_{1jk} - \frac{1}{\alpha_k} w_{jk} = 0,$$ (7)

where α_k is the waste disposal rate associated with k in tons per acre, and β_{1k} is the yield of feed i in pounds per acre from land being used for waste disposal at rate k. Four zones for waste disposal j, where $j = 1, 2, 3, 4$, and two rates of waste disposal k, where $k = 1, 2$, are assumed in this model, requiring a total of eight separate constraints within this set.

The fourth set of constraints relates to the amount of land available either for feed production or disposal of waste. This amount is assumed to be a piece-wise linear function of the radius of each zone measured to the center of the feedlot. These constraints are of the form:

$$\sum_{k=1}^{2} \frac{1}{\alpha_k} w_{jk} - K_j(y_j - R_j) = 0 \qquad (j = 1,2,3,4),$$ (8)

where Y_j is the maximum hauling distance (in miles) in zone j, R_j is the known minimum radius (in miles) of zone j, and K_j is the known rate of increase of available land area (in acres per mile) within zone j. The summation term in (8) represents the number of acres needed to spread the waste at rate α_k and must be equal to the $K_j(Y_j - R_j)$ term which is the land area available.

The fifth set of constraints sets upper and lower limits on the maximum hauling distance for waste within each zone. These are of the form:

$$Y_j \geq R_j \qquad (j = 1, 2, 3, 4) \qquad (9)$$

$$Y_j \leq R_j + D_j \qquad (j = 1, 2, 3, 4), \qquad (10)$$

where D_j is the known difference between the maximum and minimum radius (in miles) of zone j. These two constraints force the variables Y_j to remain within the limits defined for each zone in Figure 1.

The sixth and final constraint balances the total amount of annual waste spread on the land. These are:

$$\sum_{j=1}^{4} \sum_{k=1}^{2} W_{jk} - \delta Z = 0, \qquad (11)$$

where δ is the known amount of waste produced per animal per year in tons.

Additional constraints may be added to consider limits on other factors such as numbers of cattle available or amount of water used.

4. Implementation of the Model

To test the model data was obtained for feedlot operation typical of the Kansas, Oklahoma, and Texas areas. Table 1 shows the minimum daily requirements of each of the nutrients for fattening 2 year old cattle with average body weight of 900 pounds. Also listed are the amounts of each nutrient contained in each of the feeds. These values, when converted to a per pound of feed basis, are the coefficients in the first set of constraints (2).

The next four constraints (3) through (6) are to keep the silage content of the ration at least 18%, the protein supplement content at 6%, the hay content at most 12%, and the total feed consumption at most 30 pounds per day. These constraints are to ensure a ration which is attractive to the animals and meets the daily nutrient requirements within a volume that their digestive system can handle.

Table 2 shows the average yield per acre for different crops on various units and also equivalent values for acre units. The reciprocal of these values is used as the coefficient to determine the number of acres of land needed to produce the particular crop. Data collected by Kansas State University Agricultural Engineers (Ref. 5) indicates that there is a decrease in crop yield with

NUTRIENT PER POUND*

FEEDSTUFF/UNITS	Calcium %	Phosphorous %	Total Dry Matter %	Digestible Protein %	Total Digestible Nutrients %	Net Energy therms/100 lb.	Carotene mg./lb.
Alfalfa Hay	1.47	0.24	90.5	10.5	50.3	40.6	8.2
Corn Silage	0.10	0.06	27.4	1.2	18.1	18.1	5.8
Sorghum Silage	0.08	0.04	25.3	0.8	15.2	12.2	3.3
Grain Corn	0.02	0.28	87.0	6.8	82.0	82.0	1.3
Grain Sorghum	- -	- -	89.1	8.8	80.0	77.9	- -
Wheat	0.04	0.39	89.5	11.1	80.0	80.0	0.04
Cottonseed Meal (41%)	0.24	0.89	92.2	32.2	69.1	69.1	0.1
Minimum Daily Requirement**	0.044 lb.	0.044 lb.	20.7 lb.	1.53 lb.	15.0 lb.	14.0 therms	50 mg.

*These values are within the range suggested by Morrison (Ref. 5).
**Based on fattening 2 year old cattle with average 900 lb. body weight.

TABLE 1. NUTRIENT CONTENT OF CATTLE FEEDS

increased rate of animal waste disposal on the land. The two rates of waste disposal used in the model are 20 and 40 tons per acre. The average yield is used
for the fields on which 20 tons per acre are spread and a 40 percent reduction in
the average yield is assumed for the fields in which 40 tons of waste are spread.
Thus more acres of land are required to produce feed on land which has more waste
spread. The value of 0.6 has been found appropriate for all values of β_{1k}. This
value is used to determine the coefficient of each variable in the set of constraints (7) corresponding to the number of acres of land being used for feed production. These are balanced against the number of acres of land used for waste
disposal by using the appropriate coefficient, 1/20 (i.e., α_1) or 1/40 (i.e., α_2)
corresponding to the waste disposal rate, for the variables representing tons of
waste being handled.

TABLE 2. CROP YIELDS

CROP	AVERAGE YIELD/ACRE	POUNDS/ACRE
Alfalfa Hay	5 tons	10000
Corn Silage	9 tons	18000
Sorghum Silage	10 tons	20000
Shelled Corn	70 bushels	3920
Grain Sorghum	50 bushels	2800
Wheat	25 bushels	1500

*These values are based on data from Ref. 10.

The amount of land available for crop production and waste disposal is a limited resource. The total land within a given radius is equal to a constant times
the square of the radius. However, all this land is usually not available because
some may be unsuited for cropland, and some of the cropland may be ed by people
not cooperative with the feedlots. The first mile of radius is assumed to be
occupied by the feedlot itself. Zone 1 extends from 1 to 5 miles, zone 2 from 5
to 10 miles, zone 3 from 10 to 15 miles, and zone 4 from 15 to 30 miles (Fig. 1).
Figure 2 shows the assumed piecewise linear function of available land depending
on the radius. The appropriate coefficients (i.e., 1/20 or 1/40) based on the
rate of waste disposal are used for the waste variables and the slope of the line
segments in Figure 2 corresponding to each zone is the coefficient (i.e., K_j) of
the radius variable (i.e., Y_j) in constraint (8).
The set of constraints (9) and (10) place upper and lower limits on the radius
for each zone. These radius bounds (i.e., R_j and D_j) are 1 and 5, 5 and 10, 10 and 15,
and 15 and 30 for zones 1 through 4, respectively.
Beef cattle in feedlots produce approximately 60 pounds of waste daily per

Variables	Z	X_{15}	X_{25}	X_{35}	X_{311}	X_{312}	X_{321}	X_{322}	X_{331}	X_{332}	X_{341}	X_{342}
Objective Function	260.00	-.04	-.01	-.018	-.0058	-.0094	-.0065	-.01012	-.00727	-.01087	-.00877	-.01237
Set of Constraints												
Row 1	-16.06	.0024		.0147	.0147	.0147	.0147	.0147	.0147	.0147	.0147	.0147
2	-558.45	.322		.105	.105	.105	.105	.105	.105	.105	.105	.105
3	-5475.0	.691		.503	.503	.503	.503	.503	.503	.503	.503	.503
4	-16.06	.0089		.0024	.0024	.0024	.0024	.0024	.0024	.0024	.0024	.0024
5	-7555.5	.922		.905	.905	.905	.905	.905	.905	.905	.905	.905
6	-5110.0	.691		.41	.41	.41	.41	.41	.41	.41	.41	.41
7	-18250.	.10		8.2	8.2	8.2	8.2	8.2	8.2	8.2	8.2	8.2
8	-365.		1.0									
9		-.18	-.18	-.18	-.18	-.18	-.18	-.18	-.18	-.18	-.18	-.18
10		-.94	.06	.06	.06	.06	.06	.06	.06	.06	.06	.06
11		.12	.12	-.88	-.88	-.88	-.88	-.88	-.88	-.88	-.88	-.88
12	-10950.	1.0	1.0	1.0	1.0	1.0	1.0	1.0	1.0	1.0	1.0	1.0
13*					.01							
14*						.0167						
15*							.01					
16*								.0167				
17*									.01			
18*										.0167		
19*											.01	
20*												.0167
21‡												
22‡												
23‡												
24‡												
25												
26												
27												
28												
29												
30												
31												
32												
33												

*Original coefficients of constraints 13 through 20 can be obtained by dividing by 100.

‡Original coefficients of constraints 21 through 24 can be obtained by multiplying by 100.

TABLE 4. (part 1 of 4) FORMULATION OF THE MODEL

A. ASHOUR AND C. ANDERSON

Variables	X_{45}	X_{411}	X_{412}	X_{421}	X_{422}	X_{431}	X_{432}	X_{441}	X_{442}	X_{55}	X_{511}	X_{512}	X_{521}	X_{522}	X_{531}	X_{532}	X_{541}	X_{542}
Objective Function	-.0214	-.00735	-.01195	-.008	-.0126	-.0087	-.0133	-.0102	-.0148	-.018	-.0097	-.159	-.0104	-.0166	-.0111	-.0173	-.0126	-.018
Set of Constraints																		
Row 1	.0002	.0002	.0002	.0002	.0002	.0002	.0002	.0002	.0002									
2	.068	.068	.068	.068	.068	.068	.068	.068	.068	.088	.088	.088	.088	.088	.088	.088	.088	.088
3	.82	.82	.82	.82	.82	.82	.82	.82	.82	.80	.80	.80	.80	.80	.80	.80	.80	.80
4	.0028	.0028	.0028	.0028	.0028	.0028	.0028	.0028	.0028									
5	.87	.87	.87	.87	.87	.87	.87	.87	.87	.891	.891	.891	.891	.891	.891	.891	.891	.891
6	.82	.82	.82	.82	.82	.82	.82	.82	.82	.779	.779	.779	.779	.779	.779	.779	.779	.779
7	1.3	1.3	1.3	1.3	1.3	1.3	1.3	1.3	1.3									
8																		
9	-.18	-.18	-.18	-.18	-.18	-.18	-.18	-.18	-.18	-.18	-.18	-.18	-.18	-.18	-.18	-.18	-.18	-.18
10	.06	.06	.06	.06	.06	.06	.06	.06	.06	.06	.06	.06	.06	.06	.06	.06	.06	.06
11	.12	.12	.12	.12	.12	.12	.12	.12	.12	.12	.12	.12	.12	.12	.12	.12	.12	.12
12	1.0	1.0	1.0	1.0	1.0	1.0	1.0	1.0	1.0	1.0	1.0	1.0	1.0	1.0	1.0	1.0	1.0	1.0
13*		.0255																
14*			.0425															
15*				.0255														
16*					.0425													
17*						.0255												
18*							.0425											
19*								.0255										
20*									.0425									
21†											.0357	.0595						
22†													.0357	.0595				
23†															.0357	.0595		
24†																	.0357	.0595
25																		
26																		
27																		
28																		
29																		
30																		
31																		
32																		
33																		

*Original coefficients of constraints 13 through 20 can be obtained by dividing by 100.

†Original coefficients of constraints 21 through 24 can be obtained by multiplying by 100.

TABLE 4. (part 2 of 4) FORMULATION OF THE MODEL.

Variables	X_{65}	X_{611}	X_{612}	X_{621}	X_{622}	X_{631}	X_{632}	X_{641}	X_{642}	X_{711}	X_{712}	X_{721}	X_{722}	X_{731}	X_{732}	X_{741}	X_{742}	X_{811}	X_{812}	X_{821}	X_{822}	X_{831}	X_{832}	X_{841}	X_{842}
Objective Function	-.027	-.016	-.027	-.017	-.027	-.018	-.028	-.019	-.029	-.00212	-.00323	-.00279	-.0039	-.00354	-.00465	-.00504	-.00615	-.00205	-.00311	-.00272	-.00378	-.00347	-.00453	-.0049	-.0060
Set of Constraints																									
Row 1	.0004	.0004	.0004	.0004	.0004	.0004	.0004	.0004	.0004	.001	.001	.001	.001	.001	.001	.001	.001	.0008	.0008	.0008	.0008	.0008	.0008	.0008	.0008
2	.111	.111	.111	.111	.111	.111	.111	.111	.111	.012	.012	.012	.012	.012	.012	.012	.012	.008	.008	.008	.008	.008	.008	.008	.008
3	.80	.80	.80	.80	.80	.80	.80	.80	.80	.181	.181	.181	.181	.181	.181	.181	.181	.152	.152	.152	.152	.152	.152	.152	.152
4	.0039	.0039	.0039	.0039	.0039	.0039	.0039	.0039	.0039	.0006	.0006	.0006	.0006	.0006	.0006	.0006	.0006	.0004	.0004	.0004	.0004	.0004	.0004	.0004	.0004
5	.895	.895	.895	.895	.895	.895	.895	.895	.895	.274	.274	.274	.274	.274	.274	.274	.274	.253	.253	.253	.253	.253	.253	.253	.253
6	.779	.779	.779	.779	.779	.779	.779	.779	.779	.181	.181	.181	.181	.181	.181	.181	.181	.122	.122	.122	.122	.122	.122	.122	.122
7	.04	.04	.04	.04	.04	.04	.04	.04	.04	5.8	5.8	5.8	5.8	5.8	5.8	5.8	5.8	3.3	3.3	3.3	3.3	3.3	3.3	3.3	3.3
8																									
9	-.18	-.18	-.18	-.18	-.18	-.18	-.18	-.18	-.18	.82	.82	.82	.82	.82	.82	.82	.82	.82	.82	.82	.82	.82	.82	.82	.82
10	.06	.06	.06	.06	.06	.06	.06	.06	.06	.06	.06	.06	.06	.06	.06	.06	.06	.06	.06	.06	.06	.06	.06	.06	.06
11	.12	.12	.12	.12	.12	.12	.12	.12	.12	.12	.12	.12	.12	.12	.12	.12	.12	.12	.12	.12	.12	.12	.12	.12	.12
12	1.0	1.0	1.0	1.0	1.0	1.0	1.0	1.0	1.0	1.0	1.0	1.0	1.0	1.0	1.0	1.0	1.0	1.0	1.0	1.0	1.0	1.0	1.0	1.0	1.0
13*	.0667	.0667	.1111																						
14*				.0667	.1111																				
15*						.0667	.1111																		
16*								.0677	.1111																
17*										.00556	.00926														
18*												.00556	.00926												
19*														.00556	.00926										
20*																.00556	.00926								
21#																		.0050	.00834						
22#																				.0050	.00834				
23#																						.0050	.00834		
24#																								.0050	.00834
25																									
26																									
27																									
28																									
29																									
30																									
31																									
32																									
33																									

*Original coefficients of constraints 13 through 20 can be obtained by dividing by 100.

#Original coefficients of constraints 21 through 24 can be obtained by multiplying by 100.

TABLE 4. (part 3 of 4) FORMULATION OF THE MODEL.

Variables	W_{11}	W_{12}	W_{21}	W_{22}	W_{31}	W_{32}	W_{41}	W_{42}	Y_1	Y_2	Y_3	Y_4		RHS
Objective Function	-1.50	-1.50	-3.75	-3.75	-6.25	-6.25	-11.25	-11.25						
Set of Constraints														
Row 1													≥	0
2													≥	0
3													≥	0
4													≥	0
5													≥	0
6													≥	0
7													≥	0
8													≥	0
9													=	0
10													≥	0
11													=	0
12													≥	0
13*	-5.0												≤	0
14*		-2.5											=	0
15*			-5.0										=	0
16*				-2.5									=	0
17*					-5.0								=	0
18*						-2.5							=	0
19*							-5.0						=	0
20*								-2.5					=	0
21†	-.0005	-.00025							19.0				=	19.0
22†			-.0005	-.00025						100.0			=	500.0
23†					-.0005	-.00025					200.0		=	2000.0
24†							-.0005	-.00025				500.0	=	7500.0
25									1.0				≥	1.0
26									1.0				≤	5.0
27										1.0			≥	5.0
28										1.0			≤	10.0
29											1.0		≥	10.0
30											1.0		≤	15.0
31												1.0	≥	15.0
32												1.0	≤	30.0
33	≤1.0	1.0	1.0	1.0	1.0	1.0	1.0	1.0					=	0

*Original coefficients of constraints 13 through 20 can be obtained by dividing by 100.

†Original coefficients of constraints 21 through 24 can be obtained by multiplying by 100.

TABLE 4. (part 4 of 4) FORMULATION OF THE MODEL

coefficients for the waste variables (i.e., W_{jk}) in the objective function.

The variable representing hauling distance (i.e., Y_j) has a zero coefficient in the objective function, since the costs are included in the costs for the other variables.

The model, as displayed in Table 4, has been solved by IBM's Mathematical Programming System (MPS) using an IBM 360/50 computer. The maximum profit is $20,606,495.68 per year with a total feedlot capacity of 424,057.7 animals, which means about 1,272,173 animals per year. The average profit per animal is $16.19 which means that the cost of feed and waste disposal is $70.48 per animal per year or about two cents per pound of beef produced.

The ration consists of thirty pounds per day of the following ingredients:

Alfalfa Hay	12.0%
Corn Silage	27.8%
Salt	3.3%
Grain Corn	17.1%
Grain Sorghum	33.8%
Cottonseed Meal	6.0%

The land in zone 4 is not used in the optimal solution. Alfalfa hay and corn silage are produced on the waste disposal land and all other feeds are purchased from outside sources. The hay content of the ration and the total weight per animal per day are both at their upper limits.

5. *Conclusion*

The model appears to give reasonable results. The cost of producing the beef is in line with costs found in the literature. The optimal LP solution indicates that the number of cattle on feed within a 15 mile radius of the center of the feedlot is larger than any existing feedlot of comparable size in the United States. However, the industry is still developing, and may someday reach these figures in some areas. Water, which is one of the major resources considered by the feedlot industry, is not included in the model. In many areas of the High Plains, this resource may limit feedlot development below the level indicated by this model. Also, capital and supply of feeder cattle are assumed to be unlimited in the model and may, in fact, be limited in some areas. The price of cattle fluctuates quite widely and the relationship between the buying price and the selling price may affect the optimal solution significantly. These factors may be considered very easily in an extended model. Sensitivity analysis is also possible.

One problem regarding the solution is that nearly 51% of the total gain in the ration comes from outside sources which do not have to participate in the recycling. Since a major economic advantage of feedlots is in locating them in the major grain producing areas to reduce costs of transporting the feed to the cattle, this suggests that the cost of feeds from outside sources is too low. Possibly some constraint on the amount of material purchased from outside sources should be included in the model so that it better agrees with the recycling assumption.

The model indicates that recycling of wastes would be economically feasible where the feedlot management has some control over the land used for feed production. This assumption may be critical to the solution however, since all private land owners are not sympathetic to the needs of recycling.

More data is needed on the costs of handling large volumes of animal waste and the effects of these wastes on production of animal feed. This model has been tested with data of a preliminary nature only.

It is of interest to mention that grain fed cattle marketed in Kansas alone in 1967 has a gross value of over 400 million dollars. The industry is large enough so that the application of quantitative techniques would provide improvements in efficiency which would benefit both producers and consumers. The needs for protecting the quality of the environment places even more demands on management which can be assisted by using modern management techniques such as linear programming.

References

1. Dietrich, R. A.; "Cost and Economics of Size in Texas-Oklahoma Cattle Feedlot Operations," Bulletin 1083, Texas Agricultural Experiment Station, Texas A & M University, May 1969.

2. Erickson, D. B., and P. A. Phar; "Guidelines for Developing Commercial Feedlots in Kansas," Cooperative Extension Service, Kansas State University, Manhattan, Kansas, April 1970.

3. Hunt, D.; Farm Power and Machinery Management, 2nd Edition, Iowa State College Press, Ames, Iowa, 1956.

4. Johnson, W. H.; "Soil Incorporation of Beef Manure," Unpublished research report, Texas A & M University, 1970.

5. Morrison, F. B.; Feeds and Feeding, 22nd Edition, Morrison Publishing Company, Ithaca, New York, 1957.

6. Okey, R. W., R. N. Rickles, and R. B. Taylor; "Relative Economics of Animal Waste Disposal by Selected Wet and Dry Techniques," Proceedings of the Cornell University Conference on Agricultural Waste Management, January 1969.

7. Owen, T. R., and W. L. Griffin; "Research Reports: Disposal Economics Compared in Study," Feedlot Management 13 (1970).

8. Taiganides, E. P., and T. E. Hagen; "Properties of Farm Animal Excreta," Transactions of the American Society of Agricultural Engineers 9(3) (1966).

9. Wilson, C. M.; "The Plains Feeding Industry," Feedlot Management 11 (1969).

10. Agricultural Statistics - 1966, United States Government Printing Office,
    Washington, D. C., 1966.

Chapter 10

A REGIONAL PLANNING MODEL FOR THE

AGRICULTURAL SECTOR OF PORTUGAL*

Alvin C. Egbert
Central Projects Staff, Agriculture Department
Internation Bank for Reconstruction and Development
1818 H Street N. W.
Washington, D. C.

Hyung M. Kim
Central Projects Staff, Agriculture Department
Internation Bank for Reconstruction and Development
1818 H Street N. W.
Washington, D. C.

*The concepts and conclusion of this report are the authors'. The World Bank Group (IBRD, IDA, and IFC) is not responsible for the accuracy and completeness of this paper. The material presented here is part of a larger study conducted in cooperation with the Gulbenkian Foundation of Lisbon, Portugal. We are especially indebted to Messrs. F. Estacio, C. Lobao, and M. Pereira of that Foundation for their assistance and support in this work.

Abstract.   The development and validation of a planning
model for agriculture in any given region of Portugal is
discussed. The model is designed to achieve maximum
producer and consumer welfare. It answers questions as
to the kinds, amounts, and locations of products to be
produced, the investment requirements, and the production
techniques to be employed. A summary of computational
results is presented.

1. *Introduction*

Most governments are involved in general economic planning and even sectoral
planning to some degree. The degree and nature of planning usually depends on
the government's ideological base. Planning techniques range from only indirect
methods using planning instruments such as monetary and fiscal activities to a
set of detailed plans which include direct government investments in specific projects. Governments of less developed countries are likely to use more elaborate
and specific methods of planning to attempt to achieve economic and other targets.

This chapter reports on the design and validation of a planning model for agriculture in Portugal. The model is designed to provide a plan that will maximize
producer and consumer welfare. It answers questions as to the kinds, amounts, and
locations of products to be produced, the investment requirements, and the production techniques to be employed.

2. *The Model*

The basic structure of the model is outlined in the flowchart appearing in
Figure 1. The model assumes that the level of consumers' income, the cost of production – both direct and opportunity costs, the cost of transportation between
markets, and export and import prices are given. We use the following terms and
notations .

Indices
---

i $(i = 1,2,\ldots,N)$ indexes the product types.

j $(j = 1,2,\ldots,M)$ indexes the regions.

k $(k = 1,2,\ldots,K)$ indexes the types of primary production activities.

ℓ $(\ell = 1,2,\ldots,L)$ indexes the various primary agricultural products.

r $(r = 1,2,\ldots,R)$ indexes the different basic resources.

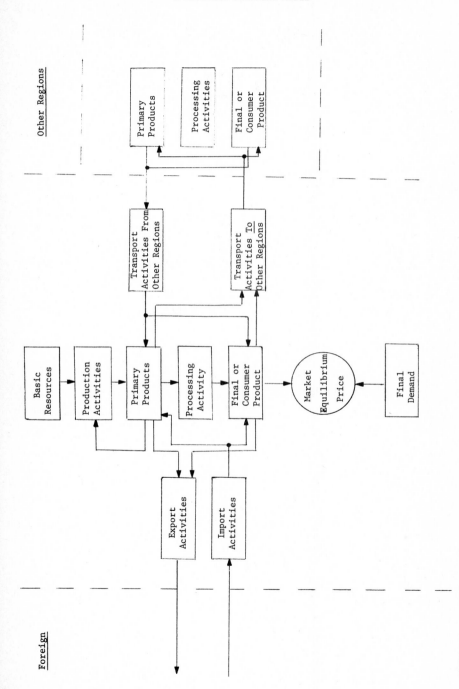

FIGURE 1.   STRUCTURE OF THE MODEL: A TYPICAL REGION

Input Parameters or Known Constants

a_{ij} The net price of product i exported* from region j.

b_{ij} The unit cost of converting a primary agricultural product ℓ into a final product in region j.

c_{kj} The unit cost of primary production activity k in region j.

d_{ij} The total import price of product i in region j.

$e_{hjj'}$ The unit transporation cost from region j to j' of the product or resource indexed by h. (h is some i, ℓ, or r.)

$\hat{I}_j$ The level of income in region j.

h_{kjr} The amount of basic resource r required for a unit of primary production activity k in region j.

$h_{\ell jr}$ The amount of basic resource r required to produce one unit of primary product ℓ in region j.

$q_{k\ell j}$ The amount of primary product ℓ produced from ($q_{kj\ell}$ will be positive) or used by ($q_{kj\ell}$ will be negative) a unit of primary production activity k in region j.

R_{jr} The supply of basic resource r in region j.

$q'_{i\ell j}$ The amount of product i produced from one unit of primary product ℓ in region j.

E_i The upper limit on the export of product i.

I_i The upper limit on the import of product i.

Decision Variables or Unknowns

W_{ij} The amount of product i sold in region j.

X_{ij} The amount of product i exported from region j.

$Y_{\ell j}$ The amount of primary agricultural product ℓ processed in region j.

U_{kj} The amount of primary of primary production activity k in region j. (Each production activity may produce more than one primary agricultural product.)

V_{ij} The amount of product i imported in region j.

$z_{ijj'}$ The amount of product i transported from region j to j'.

$z_{\ell jj'}$ The amount of primary agricultural product ℓ transported from region j to region j'.

$z_{rjj'}$ The amount of basic resource r transported from region j to region j'.

---

*The terms imports and exports refer to trade with other countries; shipment refers to interregional trade.

The Objective Function

The objective we use maximizes the total net social payoff (i.e., the producer and consumer welfare) due to the domestic sales plus the value of the exports minus the costs of processing production, transportation, and imports. The social payoff or welfare term is usually dependent on (or a function of) the amount of each product produced in a given region and the level of income in that region. Thus, suppose λ_{ij} is defined to be the average welfare measured in dollars derived from product i in region j. As λ_{ij} is a function of W_{ij} and $\hat{I}_j$, we may write $\lambda_{ij} = g(W_{ij}, \hat{I}_j)$. In our notation the objective function (1) is:

$$\text{Maximize } f = \underbrace{\sum_{i=1}^{N} \sum_{j=1}^{M} W_{ij}\lambda_{ij}}_{\text{welfare term}} + \underbrace{\sum_{i=1}^{N} \sum_{j=1}^{M} a_{ij}X_{ij}}_{\text{export income}}$$

$$- \underbrace{\sum_{\ell=1}^{L} \sum_{j=1}^{M} b_{\ell j}Y_{\ell j}}_{\text{processing costs}} - \underbrace{\sum_{k=1}^{K} \sum_{j=1}^{M} c_{kj}U_{kj}}_{\text{production costs}} - \underbrace{\sum_{i=1}^{N} \sum_{j=1}^{M} d_{ij}V_{ij}}_{\text{cost of imports}} \qquad (1)$$

$$\underbrace{- \sum_{\ell=1}^{L} \sum_{j=1}^{M} \sum_{\substack{j'=1 \\ j \neq j'}}^{M} e_{\ell jj'}z_{\ell jj'} - \sum_{r=1}^{R} \sum_{j=1}^{M} \sum_{\substack{j'=1 \\ j \neq j'}}^{M} e_{rjj'}z_{rjj'} - \sum_{i=1}^{N} \sum_{j=1}^{M} \sum_{\substack{j'=1 \\ j \neq j'}}^{M} e_{ijj'}z_{ijj'}}_{\text{transportation costs}}$$

Constraints

There are several resource, market balance, and other conditions. The first requirement ensures that for each region j and resource r the total supply (R_{jr}) added to the amount shipped in ($\sum_{j'=1}^{M} z_{j'j}$, $j' \neq j$) is not exceeded by the amount shipped out ($\sum_{j'=1}^{M} z_{jj'}$, $j' \neq j$). Or, in our notation

$$\sum_{k=1}^{K} h_{kjr}U_{kj} + \sum_{\ell=1}^{L} h_{\ell jr}Y_{\ell j} + \sum_{\substack{j'=1 \\ j' \neq j}}^{M} z_{rjj'} - \sum_{\substack{j'=1 \\ j' \neq j}}^{M} z_{rj'j} \leq R_{jr} \qquad \text{for each } j,r. \qquad (2)$$

The next constraint guarantees that for each primary product ℓ processed in region j the amount processed ($Y_{\ell j}$) is bounded above by the net amount produced added to the amount shipped in minus the amount shipped out. Therefore, we have

$$\sum_{k=1}^{K} g_{k\ell j} U_{kj} + \sum_{\substack{j'=1\\j'\neq j}}^{M} z_{\ell j'j} - \sum_{\substack{j'=1\\j'\neq j}}^{M} z_{\ell jj'} \leq Y_{\ell j} \qquad \text{for each } \ell,j. \qquad (3)$$

Inequality (4) requires that for each product i in region j the amount sold (W_{ij}) is not exceeded by the amount processed added to the amounts shipped in and imported reduced by the quantities shipped out and exported.

$$q'_{i\ell j} Y_{\ell j} + \sum_{\substack{j'=1\\j'\neq j}}^{M} z_{ij'j} + V_{ij} - \sum_{\substack{j'=1\\j'\neq j}}^{M} z_{ijj'} - X_{ij} \geq W_{ij} \qquad \text{for each } i,j. \qquad (4)$$

The next constraint guarantees that for each product i, the amount exported does not exceed the supply (E_i). That is,

$$\sum_{j=1}^{M} X_{ij} \leq E_i \qquad \text{for each } i. \qquad (5)$$

Similarly, we have that for each product i, the amount imported does not exceed the upper bound (I_i). Or,

$$\sum_{j=1}^{M} V_{ij} \leq I_i \qquad \text{for each } i. \qquad (6)$$

Finally, all variables are required to be nonnegative.

$$U_{kj}, V_{ij}, W_{ij}, X_{ij}, Y_{\ell j}, z_{rjj'}, z_{\ell jj'}, z_{ijj'}, \geq 0 \qquad \text{for every } i,j,k,\ell. \qquad (7)$$

The system (1) through (7) is a mathematical programming problem with linear constraints and a nonlinear objective function. In particular the $\sum_{i=1}^{N} \sum_{j=1}^{M} W_{ij} \lambda_{ij}$ term in the objective function is a separable nonlinear function of W_{ij}. The function λ_{ij} is known exogenously.

The nonlinear function can be approximated by a grid linearization process as described in Lasdon [9], and Hadley [7]. We will assume that this is done and the problem is solved by linear programming techniques.

3. *Data Collection*

The specific model for Portugal includes most of the major agricultural products produced and consumed in that country. These products are now identified in the following section together with a description of the sources of data and the structure of the activities.

Delineation of Regions

The first step in implementing a model of this nature is to delineate or define regional boundaries. Considerable effort went into defining regions in order
to make them as homogenous as possible. In addition, to keep the size of the
linear program managable, we required that the total number of regions should be
between 10 and 15. Portugal is a small territory covering only 8.9 million hectares (1 hectare = 2.45 acres). Thus, if 10 regions are of the same size, each
would encompass less than a million hectares each.

Regional boundaries were set after analyzing soil and topographic maps and 15
years of production data. The data used included proportion and trends in crops
produced, and trends in types and productivity of livestock enterprises followed.
The data was assembled by studying 273 concelhos (small administrative areas
similar to U.S.A. counties). Eleven regions were delineated as a result of this
analysis (see Figure 2).

Availability of Basic Resources (R_{jr})

Land. The availability of land was based on data from the 1968 Census of Agriculture which included information on total use by crops and amounts irrigated. All
land was divided into two classes, A and B. Class A land represented the top
three land capability classes of land in Portugal and encompassed about 10% of
the total agricultural land including irrigated land.

Labor Force. The labor force for the base period 1968 was estimated by interpol-
ating between census years 1960 and 1970. Since 1970 summaries were available
only by districts, it was necessary to use 1960 concelho data to allocate district
labor forces to regions in cases where regions did not include complete districts.

The labor force constraint for each quarter was expressed in adult male
equivalents using the following coefficients: Males 16-65 years, 1 unit; females
16-65 years, 0.75 units; males 12-16 years, 0.75 units; and females 12-16 years,
0.375 units. Further, the farm work week was assumed to be 45 hours or 5 days
at 9 hours per day.

Animal Power. Animal power supplies were estimated based on livestock inventories
recorded by the 1968 Census of Agriculture. The unit of measurement was hours of
equivalent power available by each quarter. Four types of animals are used for
power in Portugal. They are mules, bulls, milk cows, and beef cows. These alternative sources were standardized on the basis of their relative productivity which
was based on availability for work, size, and speed. The standard weights on
these criteria are:

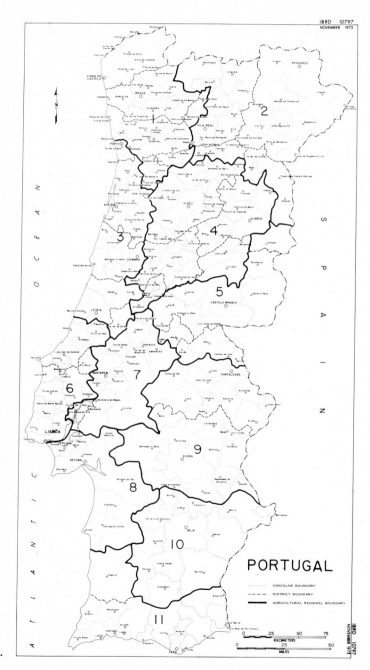

FIGURE 2.

Work bulls	1.00
Mules	.82
Beef work cows	.48
Milk work cows	.23

It was assumed that each standard animal would be available for 2,500 hours per year, or 625 hours per quarter, as availability was expressed in terms of quarters of the year.

Tractor Power. Tractor power was estimated on the basis of tractor inventories recorded in the 1968 Census of Agriculture. This resource was also specified for each quarter of the year and measured for programming purposes in terms of hours, which correspond to the unit of input for the production activities. Tractor hours were all expressed in 40 horsepower tractor equivalents. It was assumed that the standard tractor could be used 1,200 hours per year.

Processing Plants. Several types of processing constraints (capacities) were used, namely, slaughter capacity for cattle, hogs and sheep, and processing capacity for fluid milk, butter and cheese, and cold storage capacity for meat. Processing capacity for wool was not considered as a constraint because capacity is known to be more adequate for present production.

Processing capacities for livestock and products were obtained from the Junta do Pecuarios (Livestock Council) which had a special census taken in support of its own study of slaughterhouse requirements in Portugal. At present each concelho in Portugal is required by law to have its own slaughterhouse and its own processing plant.

Fertilizer Plants. Use of nitrogen and phosphorous were constrained by existing production capacity. However, there are no potash fertilizer plants in Portugal and all potash is imported through the ports of Lisbon and Porto. Because nitrogen and phosphorous fertilizer plants are located in certain towns or cities, each region does not have its own fertilizer supply. In the model, regions without fertilizer plants were supplied fertilizer through fertilizer transport activities.

Summary of Resource Constraints

        A.   Land (hectares)

            (1) Class A - irrigated   (4) Olive trees

            (2) Class A - nonirrigated   (5) Orange trees

            (3) Vineyards   (6) Other land

        B.   Labor - available man-hours by quarter.

        C.   Tractor power - available hours by quarter.

        D.   Slaughterhouse capacity - tons in higher requirement months.

        E.   Milk processing capacity - tons in critical months.

F. Butter and cheese processing capacity - tons in critical months.

G. Meat cold storage capacity - tons, cumulative requirements based
    on slaughter and consumption patterns.

H. Fertilizer plant capacity - tons per year.

Primary Production Activity: Crops (U_{kj})

All crop activities have a one year time frame and were defined on a hectare
basis. In most cases, each hectare was assumed to produce several crops in proportions representing typical rotations. For example, a typical rotation might
be corn and wheat. Thus, the output of the activity is the yield on one-half
hectare of corn and one-half hectare of wheat. In general, the rotations were for
longer periods and included more crops. Production activities for grapes, oranges,
and olives were not specified as rotation activities in the model, even though the
general practice may be to produce other crops between rows of these crops in some
regions of the country. The apparent contradiction is handled by accounting for
the land between rows of grapes or trees as part of the land base for other crops.

Crop rotations and single crop activities are differentiated by the following
factors:

A. Type of land used: Class A irrigated, class A non-irrigated, and other land;

B. System of production: Tractor power only, tractor and animal power, or ani-
    mal power only.

Thus, at the most, each crop or rotation could be represented by nine possible
activities. In practice this was not the case because certain combinations of
these alternatives were not feasible. For example, rice activities always require irrigation.

The crop production activities included in the model are:

A. Rotation crops: Wheat, corn, rice, rye, barley, oats, dry beans, broad beans,
    chick peas, potatoes, tomatoes, and forage.

B. Non-rotation crops: Grapes, olives, and oranges.

Crop production activities used the following inputs as resources:

Land. One or more of the six classes listed previously.

Labor. This input was differentiated by four seasons of the year, namely, the
standard annual quarters of January-March, April-June, July-September, and Octo-
ber-December. These monthly groupings fit the critical labor periods in the
country quite well. It would have been desirable to have specified labor requirements by months or smaller time units, but again, this procedure enlarges the
model appreciably and may produce a small increase in precision.

Tractor power. The resource tractor power is used to represent the tractor *and*
a set of machinery in the model. Other machinery in the set usually include a

plow, disk, harrow, planter, and harvester. The solution to the problem will be
the same as the one obtained when each type of machinery is a separate constraint
if each type of equipment has fixed ratio to tractors and tractors are the most
constraining piece of equipment.

Estimated hours of tractor use per rotation included time spent on tillage,
harvesting, and also hauling operations, when it is the usual practice to haul
the produce to the village market or crops to a central location for feeding.
Operational costs for tractors and machinery, which include fuel, oil, and repair
costs, were included in the objective function.

Animal Power and Machinery. These inputs were handled in a way similar to tractors. No explicit costs were attached to the cropping activity for animal power.
The costs, labor, and feed inputs, are accounted for directly or indirectly in
the objective function. It was not possible to obtain a satisfactory cost for
veterinarian services and medicines.

Fertilizer. Fertilizer application levels were those used on the average by
farmers employing the rotations considered. Fertilizer inputs were costed at
fixed government prices and included in the objective function. Although analytically desirable, costs reflecting the control of pests and plant diseases were not
included in the model because of a lack of data.

Primary Production Activities: Livestock (U_{kj})

The livestock production activities in the model are: Animal power (mules and
bulls), beef cattle, dairy cattle, swine, and sheep. All livestock activities except those producing only power (mules and bulls) were based on a livestock production unit, with the female as the base. Each livestock production unit (activity) included the adult females, adult males, replacement stock (male or female),
and male and female stock for slaughter. These were in the proportion they are
found in the average herd. Thus, a typical livestock production activity might
include ten adult females, one adult male, three female calves, three male calves,
three yearling females, and one yearling male.

Labor inputs, as with crop activities, were defined by quarters. Both crop
and livestock activities use the same labor pools or constraints. Thus, all labor
is presumed to have skill for both crop and livestock production.

Three classes of feed inputs were specified for livestock production, namely,
forage from grass, hay and silage, high protein concentrates, and low protein
concentrates. All these feeds are expressed in standard European feed units. The
low protein concentrates are derived from the feed grains - corn, oats, and barley - and high protein concentrates are derived from soybeans, peanuts, and cotton
seed; all of which are imported.

The final demand for agricultural products is measured at the retail market.
This is done principally because it is only at the retail or final consumer level
that a full set of prices for agricultural commodities are available. Also, we
want to analyze constraints on development that are related to processing facilities. This is particularly true for livestock products. There has been much
discussion in the country about the lack of processing facilities constraining the
growth of livestock production. Processing of livestock accounts for a large part
of the cost margin between the producer and consumer.

Primary Agricultural Products ($Y_{\ell j}$)

Processing activities in the model transform farm produce into consumable
forms found in retail stores. Because many products are consumed in different
forms, i.e., raw, canned, etc., these activities represent composites of the several different final products. The cost associated with these activities include
not only the cost of processing but also handling and other marketing costs except
transportation.

Domestic demand for agricultural products. A great amount of effort went into the
study of retail demand relating price and per capita consumption. Lack of regional data on consumption presented a very serious problem in estimating regional demands. A first attempt was made by estimating demand functions for each commodity
at the national level. These functions were then used to estimate quantities consumed region by region on the basis of per capita incomes and prices. The resulting consumption patterns were then compared on the basis of *a priori* expectations.
These did not turn out very well. Consequently, consumption patterns were determined on the basis of local production patterns for crops and livestock, and a
typical per capita food ration per annum was derived for each region. These
rations were then examined in terms of average kilograms, calories, and protein
consumed per person. Whenever the amounts of these items were unreasonable from a
nutritional viewpoint, national totals were reallocated so as to move regional
consumption rates into a plausible range.

After these analyses we still needed some formal expression of regional demands. Therefore, we analyzed (fit demand functions to) regional consumption patterns on the assumption that given the present patterns of consumption, the price
and income elasticities were the same in all regions. This is not very satisfactory, of course, but seemed to be the most practical course under the circumstances.

External Trade (x_{ij}, v_{ij})

Only traditionally traded commodities were represented by import or export
activities because aggregate analyses demonstrated that under the current production and demand growth only a few commodities offered real potential for export
growth. Export possibilities were wine, olive oil, lamb for special markets in
Europe, and the newly developed tomato products. On the import side, beef, pork,
wheat, rice, corn, and feed concentrates appeared to have the greatest import potential. Products that are not produced in Portugal, such as cotton, sugar, and
tobacco, and which are, by law, the special domain of the overseas provinces, were
not included. Though it would have been desirable to have export demand functions
and import supply functions such as those defined for internal markets, these could
not be estimated from available data. It is very difficult to obtain relevant export and import prices. For most agricultural imports – wheat, corn, feed concentrates, beef, and pork – it is quite realistic to assume that the import price to
Portugal does not change as the amount imported increases. This is, however, not
the case for exports of wine, olive oil, and tomatoes. The prices of these products would need to be lowered as more and more is exported. To ensure that exports would not be greater than could be marketed at the assumed export prices, a
maximum limit was placed on each of these exports. Moreover, in order to simulate
the market structure in 1968, import constraints were also imposed on wheat, corn,
beef, and pork.*

Transportation Activities ($z_{ijj'}$)

In most cases, it was assumed that agricultural products would be transported
in raw form and then processed in the region where they were consumed. This is
consistent with conditions in Portgual. Most of the products flow to or near the
large urban areas around Lisbon and Porto for processing and are consumed there.
Product exceptions in the analysis were cheese and butter which could be transported from Regions 6 to 7 and 8, and from Region 3 to Regions 1 and 4.

Most of the transport activities assumed that the hauling would be by truck.
The cost for this type of conveyance was based on a study by a German consultant
group – Planugruppe, Ribber of Konigstein, Germany (Ref. 10). Transport of grains
and fertilizer, which receive special rates, could only be transported by rail.

4. Results and Interpretations

The final matrix representing the problem we solved for Portugal consisted of

---

*The government exercises close control over all imports and exports.

1,014 rows and 4,882 columns. It was solved on a CDC 6600 computer using the
OPHELIE linear programming package at a cost of about $500 per run.

Naturally, for a planning model to be used it must be accepted by both government policy makers and planners as a valid tool. Thus, it is necessary to demonstrate to them that the model performs well. This is sometimes called the validation process. The rest of this section is devoted to validation analysis and
some discussion of how certain types of data produced by the model could be used
by planners and policy makers. A good validation, to a certain extent, depends
on the use to be made of the model. Here, we believe that for a planning model
to be useful it must reproduce, to a logical degree, the current situation. In
other words, the planner must have some idea of the starting point or base position in terms of the model's structure and capabilities. Moreover, the model must
provide him with a means for achieving the country's planning goals.

For the first part of the validation, we determine how well the model simulates recent economic activity* in production, consumption, prices, and trade.
In the second part we determine if the results behave reasonably as certain changes
are made in the model. There are three variants of this part or four in total.

The unique characteristics of the four variants of the model are:

Variant A1
A. Values in objective function
    1. Utility as given by the demand function
    2. Costs for transportation, fertilizer, tractor operations, and
        processing, handling, and selling.
B. National Constraints
    1. Exports of wine, olive oil, lamb, and tomatoes.
    2. Imports of wheat, corn, rice, beef, pork, and butter.

Variant A2
Same as A1 except there are no constraints on imports of wheat, corn, rice,
beef, and pork.

Variant A3
Same as A2 except the value of imports cannot exceed the value of exports.
(Foreign exchange balance policy.)

Variant B
Same as A3 except fertilizer and tractor operational costs are not included
in the objective function.

Variant A1 is structured to represent as nearly as possible the market conditions for agriculture in 1968 and government programs which closely manage agricultural imports.

---

*The year 1968 is used as a base because the most recent agricultural census was
taken in that year.

Variant A2 is designed to simulate what effects changes in import restriction might have on production, prices, consumption, and agricultural incomes. Variant A3 analyzes the import policy of restricting the total value of agricultural imports to the total value of exports.

Variant B examines the influence of variable production costs on output as simulated by the model. Its purpose is to give some indication of what effects changes in prices of inputs might have on output and to produce some guidelines for government price policy.

Table 1 presents the commodity production specified by optimum solutions to the 4 variants of the model together with data for the base period. Because variant A1 better represents the market and policy structure of 1968, the correspondence or correlation between it and the actual production is closer. The average correspondence is measured by the "U" coefficient.*

Commodity	Actual Production**	Variant A1	Variant A2	Variant A3	Variant B
Wheat	468.0	463.1	262.3	262.3	282.5
Rye	151.3	129.1	164.3	158.6	148.9
Corn	554.4	637.0	670.0	689.0	684.0
Rice	150.6	208.6	229.0	229.0	234.6
Barley	49.7	36.6	109.8	107.6	102.8
Oats	80.0	38.0	21.7	20.8	20.5
Broad beans	24.1	24.9	19.5	19.3	25.0
Chick peas	21.6	10.2	8.9	7.7	7.8
Potatoes	924.0	906.0	906.0	906.0	906.0
Dry beans	48.0	45.1	52.7	52.3	53.6
Oranges	106.9	106.9	106.9	106.9	106.9
Olive oil	63.6	55.6	59.2	59.2	59.2
Wine	1,162.9	1,162.9	1,162.9	1,162.9	1,162.9
Beef	61.0	50.5	50.5	51.1	51.5
Pork	105.3	104.7	104.5	104.7	104.7
Lamb and mutton	27.3	23.4	25.2	24.9	25.1
Cow milk	398.4	364.8	354.7	357.7	364.8
Sheep milk	104.5	67.0	75.3	74.2	75.3
Wool	14.1	10.4	11.3	11.2	11.3
U	-	0.085	0.137	0.134	0.133

**Normalized by using linear regression for the period 1953-67. All units are in thousands of tons or millions of liters.

TABLE 1. COMPARISON OF ACTUAL AND MODEL RESULTS

*The statistic U is measured by $U = \sqrt{\sum_i (P_i - A_i)^2 / n} \Big/ \sqrt{\sum_i P_i^2 / n + \sum_i A_i^2 / n}$, where P_i is the estimate or forecast, A_i is the actual value, and n is the number of observations. The range of U varies between 0 (perfect correspondence) and 1 (no correspondence). For a discussion of the nature and distribution of this coefficient, see Theil (Ref. 16, pp. 31-42).

The coefficient value of 0.085 roughly means that the *average difference* between actual production of all the commodities listed in Table 1, and the production specified by the solution to variant A1, is 8.5%. This result, we believe, is quite good, given the multitude of decisions in the agricultural sector that the model attempts to simulate or represent.

Under variant A2, some significant changes in the production of grain crops occur. Wheat production declines while production of rye, corn, and barley increase. In essence, these three grains use the land which was planted to wheat in the variant A1 results. This result simply means that under the production alternative specified, it is more *profitable* to import more wheat and to use the land saved to produce feed grains instead of importing them. Changes in imports and exports, as well as the total import-export picture, are given in Table 2. Wheat imports rise from 175,000 metric tons for A1 to 454,000 metric tons for A2. Contrarily, corn imports decline from 119,000 tons to 17,000.

		Imports						
		(in thousands of tons)				High Protein Feeds	Potash Fertilizer	Total value (millions of escudos)
	Wheat	Corn	Beef	Pork	Butter			
Base data	174.5	431.3	20.4	3.3	3.5*	69.0	30.0	1,527
Variant A1	164.5*	118.6*	20.4*	3.3*	3.5*	128.2	31.7	1,746
Variant A2	454.0	17.1	59.3	38.1	3.5*	127.1	37.0	3,933
Variant A3	442.0	0.2	58.7	33.0	3.5*	127.7	36.3	3,737
Variant B	437.8	0.0	58.3	34.0	3.5*	128.1	35.5	3,737

			Exports		
	Wine	Olive Oil	Tomato Products	Lamb	Total value (millions of escudos)
Base data	375.3*	21.9*	138.0*	1.1*	3,737
Variant A1			– same as base data –		
Variant A2			– same as base data –		
Variant A3			– same as base data –		
Variant B			– same as base data –		
*At upper bound					

TABLE 2. IMPORTS AND EXPORTS COMPARED WITH BASE DATA

Even though beef and pork imports increase under A2, these increases have no impact on domestic production. There is, however, significant influence on domestic consumption and prices. Beef and pork consumption increase by 38 and 35 thousand tons, respectively (Table 3), while their retail prices fall by 48% and 37% (Table 4). This result is of great benefit to the consumer, but has an unfavorable effect on producers. Producers continue to supply about the same level of output as under variant A1 because other opportunities to use their fixed resources are less profitable than beef and pork even with lower prices. The price of most crops decline under variant A2, but while the price of meat products decline, dairy products (milk, cheese, and butter) remain the same for both variants (Table 4).

Commodity	1968 Actual	Variant A1	Variant A2	Variant A3	Variant B
Wheat	580	545	664	664	664
Rye	118	98	124	121	113
Corn	225	238	239	238	239
Rice	131	146	160	160	164
Barley	17	2	2	2	2
Broad beans	35	25	19	19	25
Chick peas	15	10	9	8	8
Potatoes	818	906	906	906	906
Dry beans	55	45	53	52	54
Oranges	120	107	107	107	107
Olive oil	47	34	37	37	37
Wine	1,042	788	788	788	788
Beef	76	71	109	110	110
Pork	98	108	143	138	139
Mutton	29	22	24	24	24
Milk	322	193	201	201	211
Cheese	23	21	21	21	21
Butter	6	5	5	5	5
U	–	0.11	0.11	0.11	0.11

TABLE 3.   COMPARISON OF ACTUAL AND MODEL RESULTS ON CONSUMPTIONS
(In thousands of tons for all regions.)

Agricultural income, under variant A1, totals 13.8 billion escudos.* Under
variant A2, total agricultural income declines by over 3.1 billion escudos as
domestic farm and retail prices fall due to larger imports. On the other hand, the
net social payoff from the added consumption is 0.9 billion escudos. But the loss
of income to the agricultural sector represents a direct transfer of benefits to
consumers. Thus, the total benefit to consumers that results from removing import
constraints on wheat, corn, beef, and pork, i.e., variant A2, is 4.0 billion
escudos.

Commodity	Actual	Variant A1	Variant A2	Variant A3	Variant B
Wheat	6.00	7.08	3.43	3.45	3.43
Rye	5.17	6.03	4.87	5.06	5.35
Corn	4.95	3.99	3.99	4.05	3.96
Rice	5.89	5.43	4.96	4.96	4.83
Barley	5.93	5.20	5.19	5.34	5.10
Broad beans	8.69	9.10	9.29	9.21	9.04
Chick peas	8.20	10.12	10.12	10.96	10.91
Potatoes	1.85	1.46	1.46	1.46	1.46
Dry beans	9.67	10.87	9.44	9.40	9.39
Oranges	6.34	7.41	7.41	7.41	7.41
Olive oil**	15.73	22.10	20.11	20.11	20.11
Wine	4.54	5.83	5.83	5.83	5.83
Beef	30.60	32.60	17.13	17.13	17.13
Pork	33.22	30.90	22.48	23.53	23.32
Mutton	32.50	40.54	38.83	39.26	38.83
Milk	3.20	4.26	4.27	4.27	4.23
Cheese	56.10	61.43	61.43	61.43	61.43
Butter	42.37	46.70	46.70	46.70	46.70
Theil	–	0.11	0.11	0.11	0.11

**In escudos/liter; all other terms are in escudos/kilogram. For each commodity
the price is weighted by consumption in each region.

TABLE 4. COMPARISON OF ACTUAL AND MODEL PRICES

*At current exchange rate, one escudo is equal to $0.04.

For planning, validation of the output of the model is even more important for *regions* of a country than for country totals. It is in the regions that specific projects are implemented. Because the volume of data produced by the model is very large only some of the regional results can be shown.*

Theil U coefficients are used to measure the correspondence between regional production of each product and the specification of each variant (Table 5). The across region correspondence between base production and the model production varies significantly among products. The greatest differences are for the more or less minor crops - oats, barley, and broad beans. There is a perfect correspondence for wine because in each region all the available land in the vineyards is used by the model solution. This land produces wine equal to the 1968 base production. Regional production specification for pork is also very near to base levels in all the variants investigated because the use of breeding stock very nearly matches base inventories. Although the regional production levels are not equal to base data in every instant, the results for production, consumption, prices, and interregional trade seem sufficiently consistent to make it useful as a planning tool.

Commodity*	Variant A1	Variant A2	Variant A3	Variant B
Wheat	0.26	0.43	0.42	0.40
Rye	0.33	0.26	0.25	0.28
Corn	0.25	0.21	0.23	0.19
Rice	0.52	0.54	0.54	0.52
Oats	0.45	0.76	0.76	0.75
Barley	0.55	0.42	0.42	0.34
Broad beans	0.47	0.32	0.32	0.27
Chick Peas	0.40	0.58	0.59	0.54
Potatoes	0.31	0.37	0.38	0.39
Dry beans	0.23	0.18	0.18	0.18
Olive oil	0.12	0.08	0.08	0.08
Wine	0.00	0.00	0.00	0.00
Beef and veal	0.97	0.19	0.19	0.17
Pork	0.03	0.03	0.03	0.03
Lamb and mutton	0.11	0.12	0.12	0.12
Cow milk	0.14	0.16	0.16	0.14
Sheep milk	0.30	0.27	0.27	0.27
Wool	0.20	0.16	0.16	0.16

*There are no official farm production data by regions for livestock and livestock products. The values used for comparisons were estimated from livestock inventories.

TABLE 5. THEIL U COEFFICIENTS, COMPARING 1968 REGIONAL PRODUCTION AND PRODUCTION YIELDED BY THE MODEL

*For each of the eleven regions and each product data output of the model encompass production, consumption, processing, feed use, farm prices, retail prices, origin and destination of interregional trade, imports, and exports. In addition, there are the levels of use and surplus for each basic resource in every region.

For planning, probably, the most important set of data generated by the model are the shadow prices (or values of the dual variables) corresponding to the basic resources. The prices represent the increase in the value of the objective function that would result if one more unit of the resource was available for use in production. From a planning viewpoint, investments in a fixed resource should be made only if its shadow price is greater than its supply price. For example, under variant Al the shadow price of a hectare of orange trees turns out to be 70,000 escudos per year in Region 1. This value needs to be compared with the annual cost of establishing a grove of orange trees. Moreover, because the investment is long-term the cost should be adjusted to reflect the life of the trees.

5. *Conclusions*

When evaluating investments for development, the planner should be looking at some point in the future not a past period as we have done here. The future is where the payoffs take place. In using this model, it is best to analyze the sector for some future year, or a set of years if the structure of the sector is changing rapidly.

Our work with the model is substantially more than reported here. It leads us to believe that it has considerable potential as a planning tool in countries where (1) government economic planning is extensive, (2) the objective of planning is economic efficiency, and (3) the means for implementing plans are available.

Our collaborators at the Gulbenkian Foundation in Portugal presently are translating these results into Portuguese in the hope that planners there will see the model's merits and use it, at least to support their planning.

Even in the best circumstances the model would have to be used judiciously and experimentally, comparing the results with other data and knowledge, implementing certain prescriptions on a pilot basis, and doing repeated analysis over time as new information becomes available. Probably the model's best attribute is its power to simulate complex economic interrelationships in a framework consistent with economic theory.

References

1. Alves, Antonio Monteiro da Silva and F. Gomes; A Contribuçao de Sector Agricola Para O Desenvolvimento Economico Em Portugal, Lisboa, 1965.

2. Banco Portugues Do Atlantico; Alguns Aspectos da Economia Portguesa, Lisbon, 1966.

3. Carvalho, Cardoso and U.J. Jose; Os Solos de Portugal - Sua Classificacao e Genese.

4. Cary, F.C.; Tempos - Padroes de Trabalho Para a Cultura Arvense de Sequeiro no Alentejo, Centro de Estudos de Economia Agraria, Fundagao Calouste Gulbenkian, Lisboa, 1965.

5. Castro Caldas, Eugenio de and Santos Loureiro, Maneul de; Niveis de Desenvolvimento Agricola, No Continente Portugues, Lisboa, 1963.

6. Egbert, A.C.,and H. Kim; Analysis of Aggregation Errors in Spatial Equilibrium Programming Models, Mimeograph 1973 (submitted for publication in American Journal of Agricultural Economics).

7. Hadley, G.; Nonlinear and Dynamic Programming, Addison-Wesley Publishing Company, Inc., 1964, pp. 104-119.

8. Institute Nacional de Estatistica; Estatisticas Agricolas, Continente e Ilbas Adjacents, Lisboa, 1953-1970.

9. Lopaz, C. A.; Le Portugal, Structures Agraires et Systeme Politque, Analyse Prevision, Paris 6 6 761-780.

10. Lasdon, L.; Optimization Theory for Large Systems, Macmillan, New York, 1970, pp. 242-243.

11. Losch, A.; The Economics of Location, Yale University Press, New Haven, Connecticut, 1954.

12. Lourenco, J. Silva e Vitor Manuel Alves; Tempos de Trabalho Agricola Numa Regiao de Moroeste, Centro de Estudos de Economia Agraria Fundacao Calouste Gulbenkian, Lisboa, 1968.

13. Plannugsgrupee, R.; Portugal, Reseau National des Abattoirs: Vols. 1 - 6, Konigstun, Allemage, October 1968.

14. Samuelson, P. A.; "Spatial Price Equilibrium and Linear Programming," Amer. Econ. Rev. 42, 283-303. (1952).

15. Takayama, T. and G. G. Judge: "Spatial Equilibrium and Quadratic Programming," Jour. of Farm. Econ, 46.

16. Theil, H.; Economic Forecasts and Policy, North Holland Publishing Co., Amsterdam, 1961.

17. Yaron, D., Y. Plessner, and E. O. Heady; "Competitive Equilibrium - Application of Mathematical Programming," Can. Hour. of Agr. Econ., Vol. XIII, No. 2, 1965.

PART III

APPLICATIONS FROM OTHER ENVIRONMENTS

Chapter 11

INVESTMENT PLANNING

FOR THE TOURISM SECTOR OF A DEVELOPING COUNTRY

WITH THE AID OF LINEAR PROGRAMMING*

William W. Swart
Department of Management Science
University of Miami
Coral Gables, Florida   33124

Charles Gearing
Dartmouth College
Hanover, New Hampshire   03755

Turgut Var
University of Kansas
Lawrence, Kansas   66044

and

Gary Cann
Union Carbide Corporation
South Charleston, West Virginia   25303

*This chapter represents a modified version of an article which originally appeared in the *Tourist Review* during 1972.

Abstract. For a developing country striving to
increase its "buying power" in the community of
nations, the establishment of a tourism industry
is an important tactic for the generation of
foreign exchange earnings, and the implementation
of such a tactic frequently involves a high level
of direction from the central government. When
such governmental involvement extends to include
the direct investment of public funds in touristic facilities, then the control authority is
faced with the problem of determining the most
appropriate program for allocating the capital
investments.

This study describes an investment planning
model for the tourism sector developed for the
Turkish Ministry of Tourism. A special feature
of this model is that it directly allows the
incorporation of subjective factors associated
with the concept of "touristic attractiveness"
into the investment planning process. The model
itself is, strictly speaking, a zero-one linear
programming formulation, but it has been successfully solved using linear programming methods.
This success is attributed to the peculiar structure of a large number of the problem constraints.

The results obtained from the model provided
a basis for directing the proposed tourism development of Turkey during the third five year plan.

1. *Introduction*

For a developing country striving to increase its "buying power" in the community of nations, the establishment of a tourism industry is an important strategy
for the generation of foreign exchange earnings, and the implementation of such a
tactic frequently involves a high level of direction from the central government.
When such governmental involvement extends to include the direct investment of
public funds in touristic facilities, then the central authority is faced with
the problem of determining the most appropriate program for allocating the capital
investments. The present study constitutes an investigation of this decision
problem.

The results to be presented are a generalization of the findings obtained by
the authors while working on the specific question of "How can Turkey 'best' allocate its capital budget for tourism among a large group of competing investment
proposals for the Third Five Year Plan?" This obviously is a sub-macroeconomic
policy question and it does not consider directly the difficult higher level question of how much the touristic budget *should* be, in competition with the other

sectors. Rather, the present study takes as "given" a stipulated range within which the investment amount to be devoted to touristic projects will lie, and considers the problem of how the allocation can best be made for any planned amount within this range. The actual budgeted amount would be that amount stipulated for direct investment in touristic facilities by the Turkish government.

2. The Model

If a resource allocation can be suitably structured and quantified, it can be represented mathematically as follows:

$$\max_{X \in K} f(X) \tag{1}$$

where we have denoted the set of all feasible allocation plans by K and a typical member of that set by X, and f(X) represents the numerical measure of benefit (or "utility") associated with plan X. Thus, this mathematical statement expresses a search problem of the following sort: consider all elements of the set K and find one for which the measure of benefit is a maximum. If such a search can be effectively carried out, then a plan, denote it X*, can be found (provided it exists) which is a "best" plan (as measured by f(X)) among all the feasible allocation plans; hence, the decision problem can be resolved.

In this section we will show how a structure, in the mode of expression (1), was established for the touristic investment allocation problem. There were two main aspects to the task: (i) defining the set of feasible allocation plans (the set K above) and (ii) establishing a measure of benefit (the function f(X) above).

Defining the Set of Feasible Allocation Plans

For the purposes of the study, consider that the country, or geographical area, under consideration is subdivided into N particular touristic locations, or "touristic areas" (t.a.) and that, at any t.a., say the ith, there exist K_i specific proposed projects which may be undertaken. These projects, then represent competing investment proposals, and they cover a wide range of possible investments. Some examples are: excavation and/or restoration of ruins, construction or improvement of roads, hotels or motels, sports and recreational facilities etc. In all cases, it is expressly assumed that each t.a. has included as the first two proposed projects the following:

(1) A planning project, i.e., a proposal for a detailed development plan of

the touristic area, and

(2) A project which is designed to bring the infrastructure, food, and lodging facilities of a given t.a. up to minimally sufficient level, which was designated "minimal touristic quality" (m.t.q.).

At each t.a., the proposed projects, if undertaken, exhibit certain dependencies in the form of precedence relations derived from factors such as physical necessity, logical preference, and functional interdependence. These precedence relationships are independent between t.a.'s but, at each, the following standard convention was adopted:

(i) If a t.a. does not have a formal plan of development, the planning project precedes all others, and

(ii) If a t.a. does not have infrastructure, food, and lodging up to m.t.q. standards, the necessary improvements are considered as a single project to precede all others except the planning project.

To illustrate the above concepts, suppose that at some specific t.a. the proposed projects are designated as follows:

Description

1	Planning project
2	Infrastructure improvement to m.t.q.
3	Museum expansion
4	Thermal facilities
5	Bungalows (2000 beds)
6	Motels (1000 beds)
7	Sports facilities
8	Hotels (400 beds)

Furthermore, consider that after studying factors such as physical necessity, logical preference, and functional interdependence, the project precedence relationships are found to be as illustrated in Figure 1.

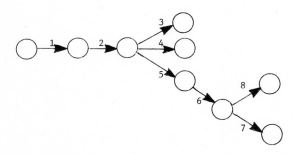

FIGURE 1. PRECEDENCE DIAGRAM FOR THE EXAMPLE

Each numbered arrow in the above diagram identifies the corresponding numbered project. The interpretation is that no project may be undertaken unless all the projects which have their arrow heads terminating on its tail are also undertaken. For example, neither project 7 or project 8 can be undertaken unless 6 is also undertaken, etc.

Associated with undertaking any project at a given t.a. is an estimated cost of completion. This cost reflects the total funding that the central planning authority must budget to that project in the event that the decision is made to undertake it, and is normally provided by initial estimates if the planning project has not been undertaken. Otherwise, the costs are reasonably accurate estimates obtained as a by-product of the planning project.

The above statements may be quantified by letting X_{ij} be a variable associated with the jth project at the ith t.a. Furthermore, let X_{ij} take on the values of 0 or 1 only. When $X_{ij}=0$ the project is *not* selected, and when $X_{ij}=1$ the the project *is* selected. With this definition of the decision variable, the precedence relationships may be expressed in terms of inequalities. In particular, the inequalities corresponding to the precedence diagram of the previous example are:

$$X_{i1} \geq X_{i2}$$
$$X_{i2} \geq X_{i3}$$
$$X_{i2} \geq X_{i4} \qquad\qquad (2)$$
$$X_{i3} \geq X_{i5}$$
$$X_{i5} \geq X_{i6}$$
$$X_{i6} \geq X_{i7}$$
$$X_{i6} \geq X_{i8}$$

In addition to the above inequalities, which must be satisfied by any feasible plan, at each t.a. there is a total cost of development which is directly related to the projects selected. For the example, the costs of completion for the various projects is given in the table below:

Project Number	Description	Estimated Cost (Turkish Lira x 10^6)
1	Planning project	0
2	Infrastructure improvement to m.t.q.	0
3	Museum expansion	5
4	Thermal facilities	20
5	Bungalows (2000 beds)	30
6	Motels (1000 beds)	30
7	Sports facilities	20
8	Hotels (400 beds)	20

The cost of projects 1 and 2 are zero in the above case because a planning study is already completed (funded from another budget), and because this particular t.a. already has an adequate infrastructure.

The total development cost, c_i, of the particular touristic area under consideration in the example may be computed through the relation

$$c_i = 0X_{i1} + 0X_{i2} + 5X_{i3} + 20X_{i4} + 30X_{i5} + 30X_{i6} + 20X_{i7} + 20X_{i8}$$

which is a function of the X_{ij}'s. In general, if we denote c_{ij} as the cost of completing project j at touristic area i, then the expression

$$\sum_{j=1}^{K_i} c_{ij} X_{ij}$$

represents the total development cost of the ith touristic area where there are K_i proposed projects, while the total development cost for all N touristic areas is given by the left side of the expression

$$\sum_{i=1}^{N} \sum_{j=1}^{K_i} c_{ij} X_{ij} \leq b. \tag{3}$$

The inequality and right side of the above expression indicates that any development policy is limited by the funds that have been made available to it. The quantity b serves to indicate this budgetary limit.

To summarize, we have indicated that a feasible touristic development plan must satisfy the precedence relationships which exist at each touristic area, and must also satisfy the budgetary limitation b. These restrictions may be expressed quantitatively as

$$\sum_{i=1}^{N} \sum_{j=1}^{K_i} c_{ij} X_{ij} \leq b \tag{4}$$

$$X_{ip} \geq X_{iq} \quad \text{for all i, and some p and q,} \tag{5}$$

where $X_{ij}=1$ implies project j at touristic area i is undertaken and $X_{ij}=0$ implies it is not. The subscripts p and q imply that project q can be undertaken only if project p is also undertaken.

Establishing a Measure of Benefit

As a country involved in nurturing the growth of tourism, it was felt that one of the primary factors to be considered during the process of developing the tourism sector of Turkey was the "attractiveness" of the developments within reach of the tourist. To incorporate this concept of "touristic attractiveness" into a formal planning model, it was required that the following be determined:

(1) The criteria by which touristic attractiveness is judged, and

(2) The relative importance of those criteria, one to the other, as indicated by a series of numerical weights.

Then, with these two requirements satisfied, it was possible to:

(3) Employ the judgement of experts in making evaluations against these criteria; and, using these inputs, to

(4) Compute a numerical measure of the "relative attractiveness" of a touristic site.

This measure of touristic attractiveness can then be used as a measure of benefit derived from investment in a particular touristic project.

The first step in establishing a measure of benefit for touristic investments is the selection of a set of "independent" criteria by which to evaluate a touristic site. The final results will depend critically upon the assumption of independence of criteria, so it is important that the selection of criteria be made to approximate this condition as closely as possible. The intuitive idea of the criteria being independent is this: regardless of how a site is evaluated on one criterion, it has no bearing on how that site would be evaluated on any of the other criteria. Of course, this is a very simple statement of a rather complex notion, discussed in detail by Fishburn (Ref. 2), but it conveys the idea in a useful, operational form; and in any application of the scope of the one being considered here, an approximation to the independence condition is the best we can expect to achieve.

The objective to be realized, in selecting the criteria, is to capture the essential ingredients of "touristic attractiveness". After considerable deliberation and study, and with close feedback from various tourism "experts", the following set of seventeen criteria were deemed to capture these essential ingredients. They are presented in five major groupings and the types of considerations involved in making valuations on each criteria are listed.

Group Heading	Criterion	Consideration
A. NATURAL FACTORS		
	1. Natural Beauty	General topography; flora and fauna; proximity to lakes, rivers, sea; islands and islets; hot and mineral water springs; caverns; waterfalls.
	2. Climate	Amount of sunshine; temperature; winds, precipitation; discomfort index.
B. SOCIAL FACTORS		
	1. Artistic and architectural features	Local architecture; mosques, monuments; art museums.
	2. Festivals	Music and dance festivals; sports events and competitions.
	3. Distinctive local features	Folk dress; folk music and dances (not organized); local cuisine; folk handicrafts; specialized products.
	4. Fairs and exhibits	Normally of commercial nature.
	5. Attitude towards	Local congeniality and treatment of tourists.
C. HISTORICAL FACTORS		
	1. Ancient ruins	The existence, condition, and accessibility of ancient ruins.
	2. Religious significance	The religious importance, in terms of present religious interests, observances, and practices.
	3. Historical prominence	The extent to which a site may be well-known due to important historical events and/or legends.
D. RECREATIONAL AND SHOPPING FACILITIES		
	1. Sports facilities	Hunting, fishing, swimming; skiing; sailing; golfing; horseback riding.
	2. Educational facilities	Archeological and ethnographic museums; zoos, botanical gardens; aquarium.
	3. Facilities conducive to health, rest, and tranquility	Mineral water spas; hot water spas, hiking trails; picnic grounds.

D. RECREATIONAL AND SHOPPING FACILITIES

	4.	Night time recreation	Gambling casinos; discoteques; theatres, cinemas.
	5.	Shopping facilities	Souvenir and gift shops; handi-craft shops; auto service faci-lities (beyond gasoline dispen-sing stations); groceries and necessities.

E. INFRASTRUCTURE, FOOD, AND SHELTER

	1.	Infrastructure above "minimal touristic quality"	Highway and roads; water, elec-tricity, and gas; safety ser-vices; health services; commu-nications; public transporta-tion facilities.
	2.	Food and lodging above "minimal touris-tic quality"	Hotels; restaurants; vacation villages; bungalows, motels; camping facilities.

Looking over this list of criteria, it is obvious that some of these criteria are more important than others. But the question to be answered is how can their relative importance be established?

The determinate of the importance of a criterion is the extent to which a tourist is responsive to the property characterized by that criterion; thus, if tourists tend to be attracted more by favorable climate than by the presence of ancient ruins, then the former is relatively more important than the latter (as far as touristic attractiveness is concerned). In order to measure the relative attractiveness, it thus seemed appropriate and expedient to use as an information source those persons who are best able to report on the behavior of tourists, namely, the tourism professionals who are in continual contact with tourists. Thus, a procedure was established for conducting an interview with a tourism "expert" for the purpose of obtaining his judgements (the procedure is an adapt-ation of one given by Churchman and Ackoff (Ref. 1)). A group of twenty-six such tourism "experts" were selected for interviews. The particular procedure was chosen as a technique for eliciting subjective judgements in a careful and con-sistent manner; and both the interviewees and the authors felt this purpose was realized in the actual application of the procedure. The results of the evalua-tions revealed that there was some basis for confidence in the particular techni-ques employed as well as statistical confirmation of the extent of agreement among the experts. A detailed discussion of the procedure and of the composition and nature of the tourism experts selected is given by Gearing, Var, and Swart (Ref.3).

The final set of weights, which were determined by simple arithmetical aver-ages of the twenty-six sets of weights, are given in Table 1, as well as the rank position of each criterion in the implied rank ordering.

	Criterion	Weight	Rank
1.	Natural Beauty	0.132	1
2.	Climate	0.099	4
3.	Artistic and architectural features	0.051	9
4.	Folk festivals	0.029	14
5.	Distinctive local features	0.026	15
6.	Fairs and exhibits	0.011	17
7.	Attitudes towards tourists	0.054	7
8.	Ancient ruins	0.057	6
9.	Religious significance	0.053	8
10.	Historical prominence	0.065	5
11.	Sports facilities	0.046	10
12.	Educational facilities	0.015	16
13.	Resting and tranquility	0.032	13
14.	Night-time recreation	0.045	11
15.	Shopping facilities	0.036	12
16.	Infrastructure above m.t.q.	0.131	2
17.	Food and lodging above m.t.q.	0.125	3

TABLE 1.   RELATIVE WEIGHTS AND RANK ORDER OF SEVENTEEN CRITERIA OF "TOURISTIC
ATTRACTIVENESS"

With the quantitative expression of the relative importance given to the various criteria of touristic attractiveness, the next step involved using these
weights is the evaluation of the various t.a.'s. To quantitatively express the
extent to which a particular t.a. satisfied the various criteria the judgements
of an "evaluation" team were sought. This evaluation team, ideally, would be composed of persons representing a number of disciplines such as tourism, architec-
ture and planning, sociology, archeology, transportation, etc. The selected team
was asked to assign, at each t.a., a valuation between zero and one on each of
the seventeen criteria. On a particular criterion, the value of zero would represent the complete absence of the property described, whereas the value of one
would represent the highest possible satisfaction of that property.

The evaluation team was charged with providing valuations on all criteria at
each t.a. when the corresponding t.a. is assumed to be in any state j. A particular t.a. is considered in state j when project j has been, or is assumed to
have been, completed along with all other projects which must precede it. Consider the column vector of valuations made at touristic area i when it is in
state j as:

$$A_i^{(j)} = \begin{vmatrix} a_{i1}^{(j)} \\ a_{i2}^{(j)} \\ a_{i3}^{(j)} \\ \cdot \\ \cdot \\ \cdot \\ a_{i17}^{(j)} \end{vmatrix} \tag{6}$$

Where $a_{ik}^{(j)}$ is the valuation of the kth criteria in touristic area i if the jth project, including its predecessors, were completed.

At each t.a. a valuation vector such as the above will exist for every project j. In addition, an initial valuation is made by the evaluation team where values are assigned to all 17 criteria which reflect the present condition (assuming m.t.q.) of the t.a. before any of the proposed projects are undertaken. As such, this valuation can be looked upon as an initial "inventory" of touristic attractiveness existing at the t.a. These valuations represent a compendium of the team members' judgements, arrived at by any number of appropriate procedures. The results of the initial valuation are recorded in the form of a column vector of values denoted as $A_i^{(2)}$.

Based on the above, we now have associated with each proposed project a beginning condition (or pre-project state) given by the valuation vector associated with the project which *immediately* precedes this project, and an ending condition (post-project state) given by the valuation vector associated with this project. Thus, the incremental change from the pre-project state to the post-project state can be represented as the vector difference of the two states, and this new vector reflects the impact of the project. This "impact vector" will be denoted by $T_i^{(j)}$ and is given by:

$$T_i^{(j)} = A_i^{(j)} - A_i^{(k)} = \begin{vmatrix} a_{i1}^{(j)} - a_{i1}^{(k)} \\ a_{i2}^{(j)} - a_{i2}^{(k)} \\ \cdot \\ \cdot \\ \cdot \\ a_{i17}^{(j)} - a_{i17}^{(k)} \end{vmatrix} = \begin{vmatrix} t_{i1}^{(j)} \\ t_{i2}^{(j)} \\ \cdot \\ \cdot \\ \cdot \\ t_{i17}^{(j)} \end{vmatrix} \tag{7}$$

Where project k is the one which immediately precedes project j in the precedence diagram.

The special conditions associated with the planning project (1) and the infrastructure improvement project (2) should be noted. It is assumed that the planning project does not, per se, affect touristic attractiveness; hence

$$A_i^{(1)} = \begin{vmatrix} 0 \\ 0 \\ 0 \\ . \\ . \\ . \\ 0 \end{vmatrix} \quad \text{at every t.a.}$$

On the other hand, the impact of the infrastructure improvement project (bringing the t.a. up to m.t.q.) is to make a non-touristic site into one suitable for
tourists, hence the impact vector reflects this fact; or

$$T_i^{(2)} = A_i^{(2)} - A_i^{(1)} = A_i^{(2)} \quad \text{at every t.a.}$$

That is, it is assumed a t.a. benefits the full measure of going from "nothing"
(touristically, that is, without adequate infrastructure, food, and shelter to
one with minimal comforts for tourists) to the initial value of "touristic attractiveness", once the minimal touristic comforts are provided for.

Given the above definition of the marginal "impact" that a particular project
will have, if undertaken, on each of the 17 criteria defining touristic attractiveness, it becomes necessary to synthesize these into a single measure of benefit for each project. Since some criteria are more important than others as
measured by the weights exhibited in Table 1, it was decided to define as the
measure of benefit for the project the weighted sum of the "marginal impacts".
In other words, the measure of benefit, d_{ij}, associated with project j at touristic area i, is given by the vector product of $T_i^{(j)}$ and W (the row vector of criteria weights ω_i); that is,

$$d_{ij} = W \, T_i^{(j)} = \omega_1 t_{i1}^{(j)} + \omega_2 t_{i2}^{(j)} + \ldots + \omega_{17} t_{i17}^{(j)}. \tag{8}$$

To illustrate the procedure presented so far in this section, consider the
example presented in an earlier section where the precedence diagram was shown
in Figure 1. All results are exhibited in Table 2.

As a first step, the initial inventory of the touristic area resulted in a

CRITERION	A_i (2)	A_i (3)	A_i (4)	A_i (5)	A_i (6)	A_i (7)	A_i (8)	T_i (2)	T_i (3)	T_i (4)	T_i (5)	T_i (6)	T_i (7)	T_i (8)	WEIGHT
1. Natural beauty	0.80	0.80	0.80	0.80	0.80	0.80	0.80	0.80	0	0	0	0	0	0	0.132
2. Climate	0.70	0.70	0.70	0.70	0.70	0.70	0.70	0.70	0	0	0	0	0	0	0.099
3. Artistic and architectural features	0.70	0.85	0.70	0.70	0.70	0.70	0.70	0.70	0.15	0	0	0	0	0	0.051
4. Folk festivals	0.70	0.70	0.70	0.70	0.70	0.70	0.70	0.70	0	0	0	0	0	0	0.029
5. Distinctive local features	0.60	0.60	0.60	0.60	0.60	0.60	0.60	0.60	0	0.10	0	0	0	0	0.026
6. Fairs and exhibits	0.60	0.60	0.60	0.60	0.60	0.60	0.60	0.60	0	0	0	0	0	0	0.011
7. Attitudes toward tourists	0.60	0.60	0.60	0.60	0.60	0.60	0.60	0.60	0	0	0	0	0	0	0.054
8. Ancient ruins	0.70	0.72	0.70	0.70	0.70	0.70	0.70	0.70	0.02	0	0	0	0	0	0.057
9. Religious significance	0.50	0.60	0.50	0.50	0.50	0.50	0.50	0.50	0.10	0	0	0	0	0	0.053
10. Historical prominance	0.70	0.70	0.70	0.70	0.70	0.70	0.70	0.70	0	0	0	0	0	0	0.065
11. Sports facilities	0.30	0.30	0.35	0.40	0.45	0.95	0.47	0.30	0	0.05	0.10	0.05	0.55	0.02	0.046
12. Educational facilities	0.30	0.40	0.30	0.30	0.30	0.30	0.30	0.30	0.10	0	0	0	0	0	0.015
13. Resting and tranquility	0.40	0.40	0.95	0.52	0.52	0.52	0.52	0.40	0	0.55	0.12	0	0	0	0.032
14. Nighttime recreation	0.30	0.30	0.40	0.40	0.45	0.45	0.70	0.30	0	0.10	0.10	0.05	0	0.25	0.045
15. Shopping facilities	0.60	0.60	0.70	0.65	0.70	0.70	0.75	0.60	0	0.10	0.05	0.05	0	0.05	0.036
16. Intrastructure above m.t.q.	0.70	0.70	0.75	0.75	0.80	0.80	0.84	0.70	0	0.05	0.05	0.05	0	0.04	0.131
17. Food and lodging above m.t.q.	0.70	0.70	0.75	0.80	0.85	0.85	0.88	0.70	0	0.05	0.10	0.05	0	0.03	0.125
d_{ij}								.90280	.01559	.04340	.03379	.09150	.02530	.02296	

TABLE 2.   ILLUSTRATIVE VALUES FOR THE EXAMPLE

valuation on all seventeen criteria as shown under the $A_i^{(2)}$ and $T_i^{(2)}$ columns.
(Since the $A_i^{(1)}$ and $T_i^{(1)}$ are always zero, they have been omitted from the table).

As a next step, new valuations are projected by expert judgement, assuming project 2 has been completed, by considering the increase on each criterion that project 3 would cause on the valuations. The new values are entered in column $A_i^{(3)}$. Note that the pre-project conditions are reflected by column $A_i^{(2)}$ and the post-project conditions by column $A_i^{(3)}$. Similarly, for all of the proposed projects, the valuations are projected by expert judgement, where the pre-project and post-project conditions are, according to Figure 1, identified as in Table 3.

Project Number	Pre-Project Condition	Post-Project Condition
3	$A_i^{(2)}$	$A_i^{(3)}$
4	$A_i^{(2)}$	$A_i^{(4)}$
5	$A_i^{(2)}$	$A_i^{(5)}$
6	$A_i^{(5)}$	$A_i^{(6)}$
7	$A_i^{(6)}$	$A_i^{(7)}$
8	$A_i^{(6)}$	$A_i^{(7)}$

TABLE 3. PROJECT VALUATIONS

To get the values for column $T_i^{(3)}$, the impact vector for project 3, the values on each criterion in column $A_i^{(2)}$ are subtracted from the corresponding values in column $A_i^{(3)}$. Similarly, for each project, the values in the 'impact vector', $T_i^{(j)}$, reflect the change in moving from the pre-project conditions to the post-project conditions. Note the two special cases (as discussed earlier) $T_i^{(1)}$ is a vector of all zeros, which indicates that the planning project has no impact (as far as "touristic attractiveness" is concerned); and $T_i^{(2)} = A_i^{(2)}$ which reflects the fact that bringing the infrastructure up to a minimal level has the impact of going from "nothing" to the valuations of the initial inventory. This latter procedure has been standardized for all sites, whether they possess minimal infrastructure or not; the distinction is introduced by designating the infrastructure improvement project cost, c_{i2}, equal to zero if the ith t.a. possess m.t.q. infrastructure.

As a last step for this touristic area, the benefit d_{i3}, associated with project 3 is found by multiplying each component of $T_i^{(3)}$ by its corresponding value in the vector of weights, W (from Table 1), and then summing. For the other projects j, the d_{ij} are found in a similar way. Note that d_{i2}, as mentioned earlier, is the weighted sum of the initial inventory values, $A_i^{(2)}$, and thereby serves as a measure of the relative attractiveness of the touristic area in its initial state (under the assumption that its infrastructure possesses minimal touristic quality).

The total benefit B_i, associated with a particular development plan for the touristic area may be computed through the relation

$$B_i = \sum_{j=1}^{K_i} d_{ij} \, X_{ij}$$

or, in our example

$$B_i = 0.0 \; X_{i1} + 0.9028 \; X_{i2} + 0.0156 \; X_{i3} + 0.0434 \; X_{i4} + 0.0374 \; X_{i5} +$$
$$0.0915 \; X_{i6} + 0.0135 \; X_{i6} + 0.0229 \; X_{i8}$$

and, in a similar fashion, the total benefit associated with a particular touristic development plan may be expressed as

$$\sum_{i=1}^{N} B_i = \sum_{i=1}^{N} \sum_{j=1}^{K_i} d_{ij} \, X_{ij}. \qquad (9)$$

Combining the results presented in the previous sections, the touristic allocation model may be stated as:

maximize
$$\sum_{i=1}^{N} \sum_{j=1}^{K_i} d_{ij} \, X_{ij} \qquad (10)$$

subject to
$$\sum_{i=1}^{N} \sum_{j=1}^{K_i} c_{ij} \, X_{ij} \leq b \qquad (11)$$

$$X_{ip} - X_{iq} \geq 0 \quad \text{for all } i, \text{ and some } p \text{ and } q \qquad (12)$$

and
$$0 \leq X_{ij} \leq 1 \quad \text{for all } i \text{ and } j. \qquad (13)$$

The expression to be maximized, (10), represents the aggregate benefit obtained from developing touristic projects over all the touristic areas. Constraints (11) represent the financial, or budgetary, limitation that must be observed when allocating funds to projects. The constraints represented by (12) serve to specify all precedence relationships that may exist between development projects at the various touristic areas.

The bounding constraints on the individual X_{ij}'s (13) have as yet not been discussed. The problem, as originally stated, required that each X_{ij} assume only values of zero or one. This requirement implies that, technically, the problem falls into a class of problems referred to as integer programming problems. Although many advances have been made in the technology available to solve such problems during the past several years, it is probably correct to say that the investment necessary to solve integer programs is greater, both from a programming standpoint and a computation expense standpoint, than that required to solve similar problems through linear programming methods. Of course, there are many instances when linear programming methods are inadequate to aid in solving integer programs.

The model presented above is a linear programming approximation to the integer program. The answers that are obtained from the linear program must be integer (i.e., each X_{ij} in the final solution must have a value of zero or one) in order for the approximation to correspond to the original problem. In the next section we will discuss how it was possible to make linear programming a valuable tool in solving the tourism investment allocation problem.

3. *Computational Results*

Before proceeding to a discussion of the computational results obtained when the model was used to help develop a tourism investment plan for Turkey, it seems appropriate to give an intuitive argument which motivated the use of linear programming to solve this problem.

Consider an extremely simple example which has only one t.a. and two projects. Assume that the formulation for that problem is

$$
\begin{array}{lll}
\text{maximize} & 3X_1 + X_2 = X_0 & \\
\text{subject to} & 8X_1 + 8X_2 \leq 14 & (11)' \\
& X_1 - X_2 \geq 0 & (12)' \\
& X_1 \leq 1 & (13)' \\
& X_2 \leq 1 & (13)'' \\
\text{and} & X_1 , X_2 \geq 0. &
\end{array}
$$

The linear programming feasible region is represented graphically in Figure 2.

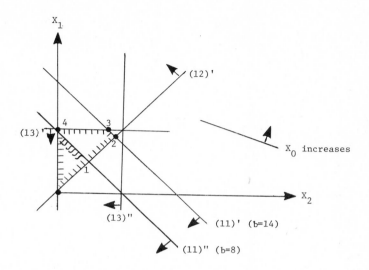

FIGURE 2.   GRAPHICAL REPRESENTATION OF THE EXAMPLE PROBLEM

As can be seen, the corner points of the region defined by the constraints (12)',
(13)', and (13)" are all integer. The constraint which creates non-integer corner
points is the budget constraint (11)'. For this situation, the optimal solution
is at corner point 3 which does not satisfy the integer requirements.

Now, consider that the amount b used for tourism planning purposes is, at the
planning stage, highly uncertain in that it represents a budget request. What
is actually received, at a later date, may be quite different from the requested
amount, hence it is not unreasonable to intepret the overall budget constraint
as representing a tentative relationship. Consequently, our reasoning was to
use parametric analysis on the budget value b to explore corner points in the
general neighborhood of the original budget constraint.

Parametric analysis is a method for investigating the effect on an optimal
linear programming solution after a sequence of proportionate changes in the
elements of a single row or column of the initial data, the budget b in our case
(Ref. 4).

Given the large number of integer corner points which usually exist in the
feasible region, it seemed plausible to expect several integer solutions to be
generated through the use of parametric analysis. In particular, this may be
illustrated by noting that a parametric analysis may, in the example exhibited
in Figure 2, shift the line (11)' to (11)". With the given objective function,
the new optimal solution is at corner point 4 and it satisfies the integer

requirements. Of course, we cannot, in general, expect that the first shift in
the line (11) will produce an integer result.(This can be seen by letting the
objective function be $1X_1 + 3X_2$ in which case the initial optimum would be at
corner point 2, and, after the first shift, it would be at corner point 1 which
is still non-integer.)

To empirically test the above ideas, a small problem consisting of 4 touristic
areas with a total of 25 projects was solved. A brief description of projects,
their cost and benefit measures, and the integer results obtained through parametric analysis, are given in Table 4. From results presented in this table
(projects undertaken are marked by an x), it was possible to obtain the typical
convex curve (Figure 3) that depicts the optimal objective function value as a
function of the right hand side value in the budget constraint. Comparing these
results to all integer solutions obtained by an integer programming scheme (at
greater cost), it was decided that linear programming with parametric analysis
would provide sufficient information for initial macro level planning purposes.

Based upon the encouraging results from the small study, we broadened the
study to include the sixty-five geographical areas which had been designated as
"touristic areas" by the Planning Department of the Turkish Ministry of Tourism.
These had been established for the purpose of planning for the development of the
tourism industry in that country. The various development projects that had been
proposed for each touristic area in the past plus projects currently in process
of being submitted were reviewed for suitability and completeness with respect to
all relevant factors. The total number of projects that remained as viable candidates for funding after this preliminary analysis was 372. The preliminary
budget figure that was proposed for funding tourism development projects for all
of Turkey over the next five year plan was $7x10^8$ Turkish Lira. Even though it
was not expected that the full amount would be granted, at least $5x10^8$ Turkish
Lira for touristic development was anticipated.

Based upon the above information, tourism planning strategies were developed
via parametric linear programming. The range within which the budget amount, b,
was allowed to vary was from $5x10^8$ T.L. to $7x10^8$ T.L. The integer solutions obtained in terms of cost and maximum benefit are listed in Table 5. Although it
is not practical to list all the corresponding solutions for this problem, as was
done in Table 4 for the small sample problem, the nature of the solutions appear
to be very similar.

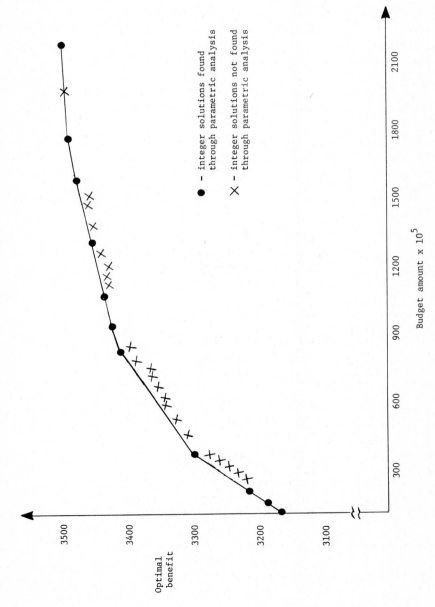

FIGURE 3. BUDGET VS. OPTIMAL BENEFIT RELATIONSHIP FOR THE SMALL STUDY

TOURISTIC AREA	PROJECT	COST ($\times 10^5$)	BENEFIT ($\times 10^{-3}$)	OPTIMAL PLANS FOR BUDGET AMOUNT [b] (in T.L. $\times 10^5$)											
				0	50	100	270	609	759	859	999	1299	1549	1749	2199
ISTANBUL	1. Planning	0	0	X	X	X	X	X	X	X	X	X	X	X	X
	2. Infrastructure to m.t.q.	0	906.0	X	X	X	X	X	X	X	X	X	X	X	X
	3. Sports facilities	200	6.0										X	X	X
	4. Recreation centers	200	3.1											X	X
	5. Thermal facilities	250	5.5							X	X	X	X	X	X
	6. Restaurants	100	14.6								X	X	X	X	X
BURSA	1. Planning	0	0	X	X	X	X	X	X	X	X	X	X	X	X
	2. Infrastructure to m.t.q.	0	760.1	X	X	X	X	X	X	X	X	X	X	X	X
	3. Airport expansion	250	19.8					X	X	X	X	X	X	X	X
	4. Recreation centers	200	22.9					X	X	X	X	X	X	X	X
	5. Camping (2000)	28	21.7					X	X	X	X	X	X	X	X
	6. Hotel (100 beds)	100	17.5								X	X	X	X	X
	7. Restaurants (150)	11	14.7					X	X	X	X	X	X	X	X
IZMIR	1. Planning	0	0	X	X	X	X	X	X	X	X	X	X	X	X
	2. Infrastructure to m.t.q.	0	924.8	X	X	X	X	X	X	X	X	X	X	X	X
	3. Camping (1000)	140	12.8							X	X	X	X	X	X
	4. Recreation facilities	100	6.3									X	X	X	X
	5. Hotel (200 first Class)	200	18.0									X	X	X	X
BODRUM	1. Planning	0	0	X	X	X	X	X	X	X	X	X	X	X	X
	2. Infrastructure to m.t.q.	0	552.6	X	X	X	X	X	X	X	X	X	X	X	X
	3. Vacation village	150	71.5					X	X	X	X	X	X	X	X
	4. Camping (1000)	20	16.4				X	X	X	X	X	X	X	X	X
	5. Vacation village (1000)	150	31.0								X	X	X	X	X
	6. Thermal facilities	50	28.0		X	X	X	X	X	X	X	X	X	X	X
	7. Motel and restaurants	50	34.7		X	X	X	X	X	X	X	X	X	X	X
Total benefit of optimal plan (original units)			3413	3178	3206	3294	3370	3401	3416	3429	3453	3473	3479	3487	

TABLE 4. SMALL STUDY RESULTS BY PARAMETRIC ANALYSIS

4. The Results, Implementation, Impact

From the results of the small study presented in Table 4, it is relatively simple to visualize the form of the results for the full scale model. Unfortunately, an extrapolation of this sort often conveys a deception as to the complexity of factors upon which the results depend. Hence, before choosing a methodology, it is essential to give consideration to some of the critical features of the analysis. In the present case the decision-maker should be alert to the pivotal nature of the subjective inputs; namely, (i) the criteria weightings derived from the twenty-six tourism experts, and (ii) the project valuations performed by the evaluation team. In case (i) there is, as reported by Gearing, Var, and Swart (Ref. 3) in their paper describing the details of the procedure, some basis for confidence in the particular techniques employed as well as the statistical confirmation of the extent of the agreement among the experts. In case (ii) the confidence of the decision-maker in the evaluation judgements (and consequently, in the final mathematical results) will be derived entirely from his confidence in the persons comprising the evaluation team. Finally, another information input which constitutes the very foundation of the analysis is the list of proposed projects. Without an adequate committment to this phase of the analysis, which would assure a reliable and realistic roster of project proposals, the validity of the entire analysis is in question. In fact, it would seem that the preparation of such a list of proposals would constitute an indispensible activity in "tourism planning".

In addition to the obvious fact of the critical nature of some of the inputs to the formal analysis, there are also some features of the methodology which constitute distinct advantages. For instance, one "by-product" is a ranking of the entire group of t.a.'s in terms of the attractiveness measure as assigned to the present condition of the t.a.'s . This ranking is provided by the computed values of $d_i^{(2)} = W A_i^{(2)}$ which represent the weighted sum of the initial inventory valuations. It can have value for purposes other than the investment allocation problem; for instance, preparing the tourism promotional program, or considering the best location for certain facilities or attractions which do not fit in with the group of local project proposals (such as fairs, sports events, cultural or educational facilities, etc.) and so on.

Another feature of the particular methodology applied in the present study is its flexibility. As an example, it might be appropriate to reconsider the decision problem from the point of view of developing mass tourism rather than individual tourism. One simple modification to accomodate this alternative perspective would be to create a new vector of weights W, which would more nearly reflect the attitudes of the purchasers of tours (or the "wholesalers" who promote them).

BUDGET AMOUNT	OPTIMAL BENEFIT FOR GIVEN BUDGET $(x10^3)*$	BUDGET AMOUNT	OPTIMAL BENEFIT FOR GIVEN BUDGET $(x10^3)*$
7000	38512.4925	5834	38011.88
6985	38507.73	5804	37996.84
6857	38453.43	5704	37946.29
6757	38427.83	5659	37923.43
6657	38391.91	5489	37835.54
6627	38380.02	5364	37770.64
6597	38367.97	5314	37744.64
6535	38341.86	5264	37718.54
6506	38329.14	5224	37697.10
6405	38284.56	5194	37680.30
6365	38266.33	5164	37663.50
6325	38248.07	5134	37646.70
6055	38119.11	5084	37618.69
6005	38094.97	5009	37576.34
5905	38046.45	5000	37571.0975
5884	38036.23		

*When the budget amount is 7000 and 5000 the solution is noninteger, and the budget constraint is at its upper and lower limit, respectively. For all other budget amounts, the solution is integer.

TABLE 5.   SOLUTIONS TO THE COMPLETE TOURIST DEVELOPMENT PROBLEM IN TURKEY

This modification could be effected by a new series of interviews, with the "experts" chosen who are knowledgeable in the area of mass tourism. A more substantial alternative of the analysis (but retaining the same methodology) would limit consideration to a series of "complex" developments designed specifically to accomodate mass tourism programs. This approach would consider the decision problem wherein the investment funds are to be allocated among competing proposals for the construction of such complexes (or, more approximately, the construction of "stages" of the complexes). It is apparent that this latter type of analysis would be devoted to a substantially different decision problem, but essentially the same methodology provides an appropriate approach to the question.

Of course there are many more questions which await recognition and answers. But the authors feel that through this study, which provides a basis for directing the proposed tourism development of Turkey during the third five year plan, a first step has been taken toward exemplifying how analytical techniques can be of use in resolving problems which arise in the management of a program of tourism development.

References

1. Churchman, C.W., and R.L. Ackoff; "An Approximate Measure of Value",
    Operations Research 2(1), 172-187 (1954).

2. Fishburn, P.C.; "Utility Theory", Management Science 14(1), 335-378 (1968).

3. Gearing, C., T. Var, and W. Swart; "Derivation of Guidelines for the
    Location of Touristic Facilities: A Turkish Example", Report prepared for
    the Turkish Ministry of Tourism, 1971.

4. Simmonard, M.; Linear Programming, Prentice Hall, Englewood Cliffs, New
    Jersey, 1966.

Chapter 12

AN APPLICATION OF LINEAR PROGRAMMING
TO THE PRELIMINARY ANALYSIS OF
RIVER BASIN PLANNING ALTERNATIVES

David H. Marks
Massachusetts Institute of Technology
Department of Civil Engineering
Cambridge, Massachusetts

and

Jared L. Cohon
The Johns Hopkins University
Department of Geography and Environmental Engineering
Baltimore, Maryland

Abstract.   The role of linear programming in the
evaluation of water resource development alternatives is considered. A complete analysis of river
basin planning problems may require several models
to capture the whole range of issues in water resource planning such as hydrologic stochasticity
and project scheduling. Although linear programming
models by themselves cannot consider all the relevant issues, they are useful for a preliminary
analysis of river basin planning alternatives.

A linear programming model is formulated for
the preliminary evaluation of alternatives in a
river basin in which national and regional development are important planning objectives. The constraint
set includes physical and technical relationships
for the various alternatives - reservoirs, power plants,
irrigation **areas, and interbasin transfers. The many**
simplifying assumptions which are necessary for the
formulation of the model are discussed. Statistics
on the size of the model and the corresponding computation costs are presented as a function of the number
of seasons and potential development sites in the
formulation.

The solutions from a formulation taken from a
real world situation which included thirteen sites
and three seasons are presented. The implications of
the results are explored and the conditions under which
an upstream interbasin export is beneficial are also
examined.

1. *The Problem*

River basin planning is the term generally used for the process by which alternatives for development in river basins are selected. Most planning problems
of this type are characterized by several alternative projects, only some of
which may be included in the final plan. Optimal development programs must be
based on an analysis of the whole river system which includes all of its components and their interactions. Thus, the river basin planning problem is, in
general, an optimization problem.

A linear programming formulation for the selection of projects in a river
basin is presented in this chapter. The program includes reservoirs and interbasin imports and exports as control alternatives, and power plants and irrigation
areas as the two primary water uses in the basin. Other water uses, such as flood
control, recreation, and navigation, are not included. To simplify the example, a
total of 13 sites are considered which have the potential for construction of six
reservoirs, five power plants, four irrigation areas, one interbasin import and
two interbasin exports associated with them.

Optimization models are potentially very useful in the analysis of water resource problems because of their ability to capture the multitude of interactions

typically observed in large systems. On the other hand, the models are limited
in application because of the assumptions we must make to formulate and solve
them. The model presented in this chapter is basically linear; water resource
systems are, however, characterized by many nonlinear relationships. To some extent these are accounted for by the use of separable programming and some formulation "tricks." However, many of the nonlinearities cannot be included in a linear
programming model. River basin systems are also characterized by a high degree of
hydrologic stochasticity, i.e., streamflows are uncertain. The present model is
deterministic in that mean seasonal streamflows are employed. This represents a
fairly strong assumption about the hydrologic processes in the river system.
Finally, the model is steady-state in that only one year is explicitly modeled
and every subsequent year up to the planning horizon (fifty years) is assumed to
be hydrologically identical to the first. This assumption precludes the adequate
consideration of two important aspects of the planning problem: optimal operating
policies for the river basin (i.e., when to store water in reservoirs during wet
periods for use in dry periods) and project scheduling (i.e., optimal construction times for projects).

All of the above considerations have led to the use of optimization models for
the *preliminary* analysis of river basin planning problems. A model of this type
is called a screening model in the water resources literature, because it is used
figuratively as a screen through which all of the possible combinations of planning alternatives are passed. Screening models are also discussed in References
1, 5, 13, 15, 19, 20 and 22.

After results are obtained from a screening model, further analysis is required to evaluate the alternatives in light of the assumptions in the screening
model. Nonlinearities, stochasticity, and operating policy are usually accounted
for in a hydrologic simulation model which simulates the operation of a given
system design for an arbitrarily long period. Simulation models are not usually
appropriate for the selection of optimal plans by themselves, but they are very
effective when used in conjunction with a screening model. Screening simulation
studies of this kind were first discussed in References 6, 8, and 18. The development of these approaches are in References 4, 11, and 14.

The screening model is not formulated in the same way for all applications.
For some problems, certain aspects of the formulation or a particular assumption
may be especially critical. In some cases, hydrologic uncertainty may be of
great concern for a particular application, so that it should be incorporated
even in preliminary analyses. Stochastic considerations in screening models are
discussed in References 7, 12, 16, and 17. It should be mentioned however, that
including stochastic considerations in the screening model or relaxing any of the
other assumptions will increase solution costs. In many cases, the increase in

computation costs may be significant to the point of being prohibitive.

2. The Model

The river basin planning problem is basically a resource allocation problem.
There are scarce resources (water and usually capital) which must be allocated
among water uses (hydroelectric energy production and irrigation) to maximize
a set of planning objectives. In addition, there are control alternatives
(reservoirs) which allow the resources to be used more effectively. The objective
function expresses the set of planning objectives in terms of the decision variables in the model, namely, decision variables representing the release of water
from reservoirs, the diversion of water out of the stream for water uses, the
realizable production from uses to which water is allocated, and the location
and capacities of the structural components of the river system.

The constraint set consists of continuity constraints which trace the flow of
water through the river system and constraints on each of the elements or uses of
the system, namely, reservoirs, irrigation, hydroelectric energy production, and
interbasin imports and exports. Each of these types of constraints are discussed
separately below, after which the objective function is presented.

Continuity Constraints

Continuity constraints are included in the model to trace the flow of water
through the river system by ensuring the conservation of mass at every point in
the river at which water is stored, diverted, or imported. A sample site with a
reservoir is shown in Figure 1.

The basic continuity relationship can be written, referring to Figure 1, as

$$V^2_{i,j+1} = V^2_{ij} + Q_{ij} + W^2_{ij} - Y^2_{ij} - X^2_{ij}, \qquad (1)$$

where the subscripts i and j refer to site and season, respectively. Equation
(1) says that the storage in the reservoir at the beginning of the next season
($V^2_{i,j+1}$) must equal the storage at the beginning of the present season (V^2_{ij}) plus
any additions during the present season (the inflow, Q_{ij} and any imports, W^2_{ij})
minus any deductions during the present season (the reservoir release, X^2_{ij} and
any diversions, Y^2_{ij}).

In equation (1), all of the variables except Q_{ij} represent decisions that are
made at site i. On the other hand, Q_{ij}, the upstream inflow, depends on natural
streamflow and on the decisions made immediately upstream, at site i-1. It is
necessary to express Q_{ij} as a function of these two effects. Figure 2 illustrates

the following development.

Applying the same conservation of mass principle as in equation (1), we get
from Figure 2 that

$$Q_{ij} = X^2_{i-1,j} + \Delta F_j,$$ \hfill (2)

where all variables are defined as before and ΔF_j represents the increment to
natural streamflow between sites i-1 and i during season j, is a known constant.
Equation (2) is substituted into (1) and after rearranging terms we have

$$V^2_{i,j+1} - V^2_{ij} + X^2_{ij} + Y^2_{ij} - W^2_{ij} - X^2_{i-1,j} = \Delta F_j,$$ \hfill (3)

in which all terms appearing on the left-hand side are decision variables while
ΔF_j on the right-hand side is input to the model. The appropriate value of ΔF_j
to use in (3) represents the major hydrologic question in the model because ΔF_j is
the only representation of natural streamflow in the constraint set.

As mentioned previously, the model presented here is deterministic in that ΔF_j
is based on mean seasonal streamflows that are taken to occur with certainty. The
value of ΔF_j is taken as the difference between the mean seasonal streamflows at
sites i-1 and i, and represents the flow into the stream from the drainage area
between i-1 and i as shown in Figure 3. Notice that ΔF_j is not the total streamflow at site i because $X^2_{i-1,j}$, which appears in (3), is the flow in the stream at
site i-1 after development of the basin. Thus, ΔF_j represents the increment (or
decrement) to $X^2_{i-1,j}$ due to natural effects, e.g., runoff from precipitation or
snowmelt, tributary flow, infiltration or exfiltration, evaporation, and any
existing developmental effects. However, none of these hydrologic processes are
treated in the model, rather ΔF_j is simply measured as the difference between
measured mean seasonal streamflows at sites i-1 and i; thus, we may write,

$$\Delta F_j = F_{ij} - F_{i-1,j},$$ \hfill (4)

where the values of F_{ij} and $F_{i-1,j}$ the mean seasonal streamflows, at sites i and
i-1, respectively, are obtained from streamflow records available at gauging
stations on the river.

With estimates of streamflows, substituting (4) into (3) results in the
stream continuity equation

$$V^2_{i,j+1} = V^2_{ij} + X^2_{ij} + Y^2_{ij} - W^2_{ij} - X^2_{i-1,j} = F_{ij} - F_{i-1,j}.$$ \hfill (5)

The storage terms, $V^2_{i,j+1}$ and V^2_{ij} are expressed in cubic hectometers per season

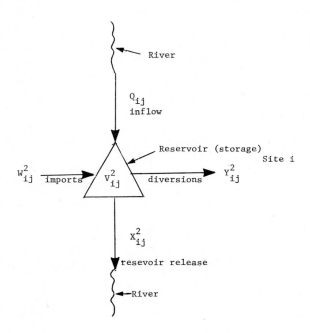

FIGURE 1. THE CONTINUITY RELATIONSHIP

(hm^3/season), where one hm^3 equals one million cubic meters (m^3). All of the other terms in (5) are average flows expressed as cubic meters per second (m^3/sec). For dimensional consistency the storage terms must be converted to m^3/sec. This is done by multiplying $V^2_{i,j+1}$ and V^2_{ij} by (1 hm^3/seas)x(1/k_j seas/secs)x(10^6 m^3/hm^3) = (10^6/k_j)x[(m^3/sec)/(hm^3/seas)], where k_j is the number of seconds in season j; multiplying this conversion factor times the storage terms in (5) gives the final form of the stream continuity constraint below.

$$\frac{10^6}{k_j} V^2_{i,j+1} - \frac{10^6}{k_j} V^2_{ij} + X^2_{ij} + Y^2_{ij} - W^2_{ij} - X^2_{i-1,j} = F_{ij} - F_{i-1,j} \quad \text{for all i,j.}$$

$$(6)^*$$

*An asterisk represents the form of the constraint that is included in the model.

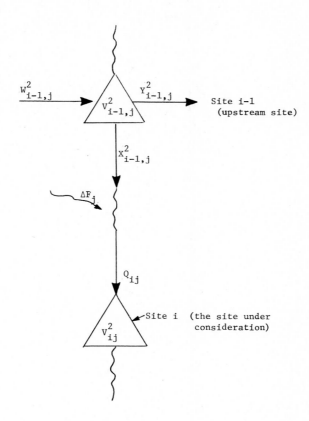

FIGURE 2. THE CONTINUITY RELATIONSHIP WITH AN UPSTREAM SITE

Note that (6) is intended to be a general continuity constraint which is written for all sites and seasons. When writing the constraint for a specific site, other terms may appear. W_{ij}^2 and Y_{ij}^2 are surrogate quantities for any inflows or outflows (other than upstream inflow and release). In an actual application these two terms may represent interbasin imports or exports, or diversion for irrigation. All of the possible forms that (6) might take were omitted here to preserve clarity of exposition. The reader should realize, however, that the form of (6) is different for each type of developmental site.

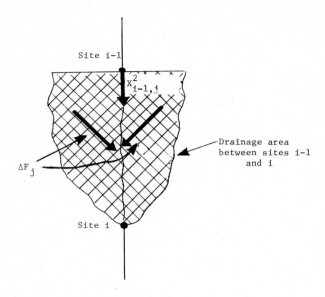

FIGURE 3.   REPRESENTATION OF NATURAL INFLOW INTO SITE i

Reservoir Constraints

There are two purely physical relationships for reservoirs. We require that the storage in a reservoir cannot exceed the storage capacity during any season j or at any site i, or

$$V_{ij}^2 - V_i \leq 0 \qquad \text{for all } i,j, \qquad (7)*$$

where V_i is the storage capacity of the reservoir at site i in hm^3. Notice that by making V_i a decision variable in the model, the optimal storage capacity of the reservoir can be found.

We also need the storage-head relationship for reasons explained in the later section on constraints for hydroelectric energy production. The constraint says simply that the storage, V_{ij}^2, is related to the height of water behind the dam in meters, x_{ij}^1, by

$$V_{ij}^2 - \sigma_i(x_{ij}^1) = 0 \qquad \text{for all } i,j, \qquad (8)*$$

where $\sigma_i(x_{ij}^1)$ is a function which relates storage volumes to the water elevation in the reservoir.

Irrigation Constraints

The irrigation process is extremely complex, and therefore quite difficult to
model by linear programming. This complexity is primarily due to the great number
of variables that affect agricultural production. Unlike hydroelectric power production, which depends only on the reservoir head and the flow through the turbines,
crop production depends on irrigation water volumes, temporal distribution of irrigation water volumes, water quality (e.g., salinity), solar radiation, precipitation, and a host of soil properties. Furthermore, the significance of each of
these variables varies from crop to crop.

What is desired in modelling an irrigation system is a production function,
i.e., a function that relates crop yield to quantities of water supplied for irrigation. The agricultural production function, which has many dimensions, one for
each of the variables which affect crop yield, has yet to be derived analytically.
Instead, the most widely used approach for estimating the production function has
been empirical investigation. By observing crop yields for varying water quantities an estimate of the production function is found. The basic weakness of this
approach is that the other variables that affect the growing process vary from one
observation to the next. However, the empirical method is the only workable technique currently available.

A number of simplifications were made in order to apply the empirical method to
this model. These are graphically shown in Figure 4. Not knowing the actual
shape of the production function, one point on the function was estimated from
empirical data (Figure 4a). Given this point, it was assumed that the annual
amount of irrigation water used at site i per hectare (10,000 square meters), τ_i,
was a constant, giving a constant yield per hectare. Therefore, as water used for
irrigation is varied the amount of land irrigated varies, not crop yield per
hectare which is constant. Furthermore, this annual land to water relationship
is linear with slope τ_i as shown in Figure 4b. The annual volume of water applied
to each hectare must be properly distributed over time to reflect the variation in
crop requirements over the year. For example, many crops need most of their annual
requirement during the summer and considerably less water during the winter.
Figure 4c shows how the annual requirement, τ_i, is distributed over a year which
has been divided into three, four-month seasons. The seasonal proportions are then
used to arrive at seasonal irrigation water use coefficients, τ_{ij}, measured in
m^3/hectare at site i during season j.

The linear function relating irrigated land U_{ij}^1 in hectares (ha.) to the volume
of water supplied for irrigation in hm^3, W_{ij}^1, is in equation (9).

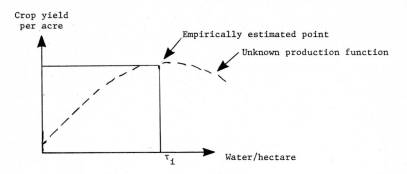

FIGURE 4a.   AGRICULTURAL PRODUCTION FUNCTION

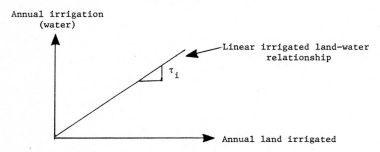

FIGURE 4b.   ASSUMED LAND-WATER RELATIONSHIP

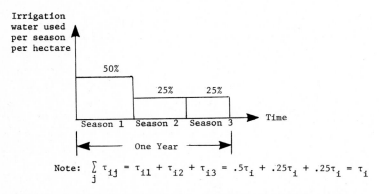

Note: $\sum_j \tau_{ij} = \tau_{i1} + \tau_{i2} + \tau_{i3} = .5\tau_i + .25\tau_i + .25\tau_i = \tau_i$

FIGURE 4c.   TEMPORAL DISTRIBUTION OF IRRIGATION WATER REQUIREMENTS

$$W^1_{ij} - \frac{\tau_{ij}}{10^6} U^1_{ij} = 0 \qquad \text{for all } i,j. \qquad (9)*$$

In general, the choice of which crops are to be produced can be a decision variable in the model. It is assumed, however, that a "cropping pattern" is chosen
prior to solving the model. The choice of crops will dictate the values of τ_{ij}
to be used in the model, and thus τ_{ij} is known under our assumption.

If benefits from irrigation are to be realized, we must be sure that land
irrigated during a season, U^1_{ij}, also receives its water requirements during other
seasons. For example, if 1000 hectares are irrigated in season 1, but only 500
ha. receive water during season 2, we cannot expect to reap benefits from the
entire 1000 ha. To discourage this type of result in the model, benefits are
related to land which receives its irrigation water requirement in every season.
This is done by computing the minimum seasonal irrigated land area, U_{im}, from the
constraints,

$$U_{im} - U^1_{ij} \leq 0 \qquad \text{for all } i,j, \qquad (10)*$$

where U^1_{ij} is the land area irrigated at site i during season j. Benefits from
agricultural production are related to U_{im} in the objective function.

The remaining irrigation constraints are derived from a consideration of
Figure 5. One constraint relates the volume of water which reaches the irrigation
site, W^1_{ij}, in hm^3 to the flow diverted out of the stream in m^3/sec, (Y^2_{ij}); this
constraint is

$$W^1_{ij} - \frac{k_j}{10^6} (1-\varepsilon_{ij}) Y^2_{ij} = 0 \qquad \text{for all } i,j, \qquad (11)*$$

where k_j is the number of seconds in season j and converts Y^2_{ij} from m^3/sec to m^3,
and ε_{ij} is a coefficient that represents the water lost in transport from the
stream to the irrigation site which is assumed to return to the stream.

Another constraint relates the flow diverted for irrigation, Y^2_{ij}, to the flow
which returns to the stream from the irrigation site in m^3/sec, V^1_{ij}; this constraint is

$$V^1_{ij} - (1-\mu_{ij}) Y^2_{ij} = 0 \qquad \text{for all } i,j, \qquad (12)*$$

where μ_{ij} is the total loss coefficient for irrigation. It represents a combination of the losses due to transport (ε_{ij}) and consumption requirements (ρ_{ij}).
As shown in Figure 5, $\mu_{ij} = \rho_{ij}(1-\varepsilon_{ij})$, when all transport losses are assumed to
return to the stream.

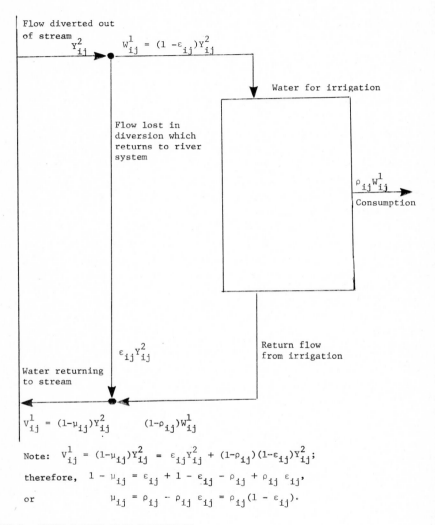

Flow diverted out of stream Y^2_{ij}

$W^1_{ij} = (1 - \varepsilon_{ij})Y^2_{ij}$

Water for irrigation

$\rho_{ij}W^1_{ij}$
Consumption

Flow lost in diversion which returns to river system

Return flow from irrigation

$\varepsilon_{ij}Y^2_{ij}$

Water returning to stream

$V^1_{ij} = (1-\mu_{ij})Y^2_{ij}$ $(1-\rho_{ij})W^1_{ij}$

Note: $V^1_{ij} = (1-\mu_{ij})Y^2_{ij} = \varepsilon_{ij}Y^2_{ij} + (1-\rho_{ij})(1-\varepsilon_{ij})Y^2_{ij}$;

therefore, $1 - \mu_{ij} = \varepsilon_{ij} + 1 - \varepsilon_{ij} - \rho_{ij} + \rho_{ij}\varepsilon_{ij}$,

or $\mu_{ij} = \rho_{ij} - \rho_{ij}\varepsilon_{ij} = \rho_{ij}(1 - \varepsilon_{ij})$.

FIGURE 5. WATER LOSSES FROM IRRIGATION

Hydroelectric Energy Constraints

The production of hydroelectric energy is a relatively well defined technical process. There are only three decision variables which affect energy production, namely, the flow through the turbines of the power plant, the head (i.e., poten-

tial energy) associated with this flow, and the capacity of the power plant. The
relationships of these variables to energy production are the origins of the
energy constraints.

The first constraint is the production function for hydroelectric energy; it
is

$$U^2_{ij} - (2.61 \times 10^{-6}) (e)(k_j) \, X^2_{ij} \, X^1_{ij} \le 0, \tag{13}$$

where U^2_{ij} is the energy produced at site i during season j in megawatt-hours
(mw-hr), X^2_{ij} is the average release from reservoir i during season j in m^3/sec,
X^1_{ij} is the head of water in reservoir i at the beginning of season j in meters,
e is the power plant efficiency, k_j is the number of seconds in season i, and
(2.61×10^{-6}) is a unit conversion factor.

Expression (13) is nonlinear because X^2_{ij} and X^1_{ij}, both decision variables, are
multiplied together. The constraint may be made linear by writing it as the two
constraints below, one with an assumed value for the release, $\hat{X}^2_{ij}$, and the other
with an assumed value for the head, $\hat{X}^1_{ij}$. The assumed values of $\hat{X}^2_{ij}$ and $\hat{X}^1_{ij}$
were computed from knowledge of the sites.

$$U^2_{ij} - (2.61 \times 10^{-6}) \ (e)(k_j)(\hat{X}^2_{ij}) \, X^1_{ij} \le 0 \qquad \text{for all i,j,} \tag{14}*$$

$$U^2_{ij} - (2.61 \times 10^{-6}) \ (e)(k_j)(\hat{X}^1_{ij}) \, X^2_{ij} \le 0 \qquad \text{for all i,j.} \tag{15}*$$

After solution of the model the assumed values, $\hat{X}^2_{ij}$ and $\hat{X}^1_{ij}$, were compared to the
computed values. If satisfactory agreement was not found, new assumed values were
used and a new solution obtained. This iterative approach was found to converge
in all cases in at most two runs. Other linearization techniques, which relied
on one constraint rather than two, with either an assumed head or an assumed
release, often gave physically unrealistic solutions.

The only other variable is the power plant capacity. The capacity represents
an obvious upper bound on energy production, namely,

$$U^2_{ij} - h_j z_i \le 0, \tag{16}$$

where h_j is the number of hours in season j, and z_i is the capacity of the power
plant i in megawatts (mw). Equation (16) will be binding only if the plant produces at capacity all of the time. This would be unrealistic and undesirable.
Therefore a load factor, p_{ij}, which is defined as the ratio of the average daily
production to the daily peak production, is introduced into (16) to represent the
daily variation in production. However, since U^2_{ij} is the seasonal energy produc-

tion it must be assumed that production does not vary appreciably from day to
day. Equation (16) becomes,

$$U_{ij}^2 - P_{ij}h_j z_i \leq 0 \qquad\qquad \text{for all } i,j, \qquad\qquad (17)*$$

where P_{ij}, an input parameter, can be between 0 and 1, and is found from assumptions based on loading histories of similar installations.

Interbasin Imports and Exports

Interbasin transfers are modeled simply as diversions of water into or out of
the stream. It is required, additionally, that the seasonal transfer does not
exceed the channel capacity. For imports, this is written as,

$$W_{ij}^2 - W_i \leq 0 \qquad\qquad \text{for all } i,j, \qquad\qquad (18)*$$

where W_{ij}^2 is the average import at site i during season j in m^3/sec, and W_i is
the capacity of the import canal at site i in m^3/sec. Similarly for exports,

$$Y_{ij}^1 - Y_i \leq 0 \qquad\qquad \text{for all } i,j, \qquad\qquad (19)*$$

where Y_{ij}^1 is the average export from site i during season j in m^3/sec, and Y_i is
the capacity of the export canal at site i in m^3/sec.

The Objective Function

The objective function for water resource screening models is, generally
multidimensional which reflects the multiobjective nature of investments in
river basin planning alternatives. The present model was formulated to account
for two objectives: national and regional development. However, for the present
discussion, only the national development objective is considered. The reader
is referred to Cohon (Ref. 2), and Cohon and Marks (Ref. 3) for a detailed discussion of the multiobjective aspects.

The national development objective is represented in the objective function
by the maximization of net national income benefits. This is identical to the
objective of economic efficiency which is traditionally used in water resource
planning. The objective function consists of gross benefits accruing from
hydroelectric energy production, irrigation, and exports to other river basins.
Subtracted from these benefits are the capital and operationg costs for reservoirs, power plants, irrigation works, and export and import canals. Mathemati-

cally the objective function is

$$\text{maximize } Z = \sum_i \left(\sum_j (\beta^P_{ij} U^2_{ij}) + \beta^I_i U_i \right) + \sum_i \sum_j \nu_{ij} Y^1_{ij} -$$

$$\sum_i \left(\alpha_i V_i + \delta_i Z_i + \phi_i U_i \right) - \sum_i \gamma^E_i Y_i - \sum_i \gamma^I_i W_i, \qquad (20)*$$

where all of the benefits and costs are discounted with an appropriate discount rate. In the equation, β^P_{ij} and β^I_i are unit benefits for power and irrigation, respectively, ν_{ij} is the unit benefit derived from interbasin export of water from site i during season j, α_i, δ_i, and ϕ_i are unit costs associated with building reservoirs, power stations, and irrigation sites, respectively, and γ^E_i, γ^I_i are unit costs of exports and imports at site i, respectively. In general, the benefit and cost functions may be nonlinear in which case they are approximated by piecewise linear functions and solved through the use of separable programming (see Ref. 21, pp. 551-554).

3. Results, Implementation

The model was applied to a river basin in South America, a schematic of which is presented in Figure 6. Two programs were solved for the configuration shown in Figure 6, one with two seasons and the other with three seasons. In addition, formulations which included more sites due to disaggregation of old sites and the incorporation of new ones were also solved. Table 1 shows the sizes of these formulations and their computational requirements.

Sites	Seasons	Constraints	Structural Variables	Iterations to Optimal Solution	Time to Optimal Solution (minutes)	Cost of Solution ($)
13	2	134	132	133	0.17	2.00*
13	3	196	187	224	0.33	6.00*
28	3	435	422	577	3.27	21.70*
38	3	629	665	800	3.90	33.00+

*Results obtained using IBM/MPS routine (Ref. 9) on an IBM 360/165 computer.
+Results obtained using IBM/MPSX routine (Ref. 10) on an IBM 370/155 computer. The iterations, time, and cost to optimal solution are approximate.

TABLE 1. COMPUTATIONAL REQUIREMENTS AS A FUNCTION OF THE NUMBER OF SITES AND SEASONS

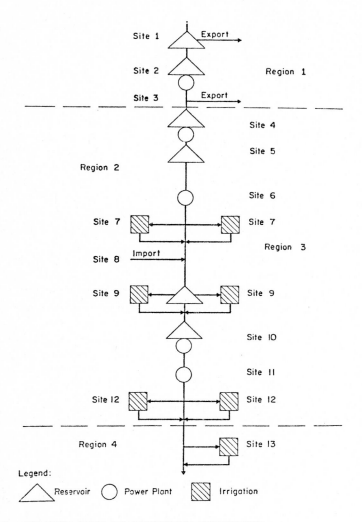

FIGURE 6.   THE RIVER BASIN AND DEVELOPMENT ALTERNATIVES

The thirteen site/three season model corresponding to the configuration in
Figure 6 was solved several times under various sets of assumptions. Of particular concern in the application was whether or not water should be exported out
of the river basin at sites 1 or 3 in Figure 6. A large export would result in
a significant loss of benefit to downstream regions (regions 2, 3, and 4 in
Figure 6). The results from one run are shown schematically in Figure 7 in

which alternatives which are shaded in are those that entered the solution.

Figure 7 is useful for understanding the operation of the system. The amount of shading for each site shows how much capacity should be built. The reservoir at site 2 is included to produce energy and to control the river for the export of water at site 3 (essentially all of the streamflow at this point). The small reservoir at site 4 is provided for energy production and control for the small irrigation site at site 7. The reservoir at site 9 is included for the sole purpose of controlling the import at site 8 to provide the appropriate seasonal flow distribution for the major downstream irrigation areas at sites 12 and 13.

The results shown in Figure 7 indicate that the export was beneficial for a given set of conditions. A primary concern was to investigate how results would change as conditions varied. Subsequent runs were made to explore the competitive relationships between the export and downstream development, since it was obvious that the export must preclude at least some of the activity in downstream sites. The analysis showed first that the export competed directly with only the sites that were downstream of the import at site 8. It was further concluded that when the import was not in the solution (it was constrained to be zero) then the export did not compete favorably with the lower basin, i.e., it did not enter at the full capacity. Alternatively, when the import was in the solution then the export was beneficial at its upper bound.

Results such as these were useful in providing the analysts and the decision makers new insight into the design of a typical regional system. Recall, however, that the screening model was intended only as a preliminary step in the design process. Some of the assumptions that were made in the model were later relaxed when the results were further analyzed in a simulation model. The conclusion of this next phase of the design process was that, in general, the screening model tended to be "optimistic," in that reservoirs were designed to be too small for the level of development in the basin. This was not unexpected, however, because the use of mean streamflows does not allow the screening model to consider the variability associated with river systems. This uncertainty was considered during the simulation phase which resulted ultimately in more reliable designs for the system.

References

1. Blanchard, B.; "A First Trial in the Optimal Design and Operation of a Water Resource System," M.S. Thesis, Dept. of Civil Engineering, Massachusetts Institute of Technology, 1964.

2. Cohon, J. L.; "Multiple Objective Screening of Water Resource Investment Alternatives," M.S. Thesis, Dept. of Civil Engineering, Massachusetts Institute of Technology, 1972.

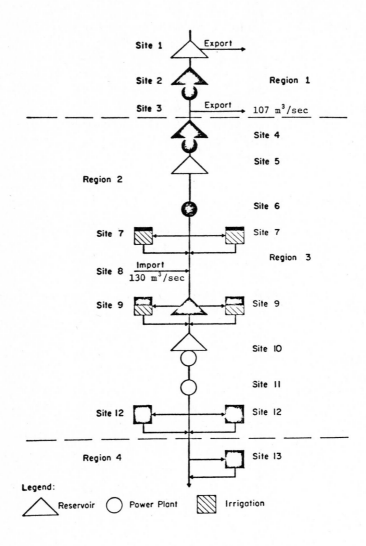

FIGURE 7.   A DESIGN FOR THE RIVER BASIN

3. Cohon, J. L., and D. H. Marks; "Multiobjective Screening Models and Water
    Resources Investment," Water Resources Research 9(3), 826-836 (1973).

4. Cohon, J. L., T. B. Facet, A. H. Haan, and D. H. Marks; "Mathematical Pro-
    gramming Models and Methodological Approaches for River Basin Planning,"
    Technical Report, Ralph M. Parsons Laboratory for Water Resources and Hydro-
    dynamics, Massachusetts Institute of Technology, forthcoming.

5. Dorfman, R.; "Mathematical Models: The Multistructure Approach," in Design
    of Water Resource Systems (ed.: A. Maass, et. al.), Harvard University Press,
    Cambridge, Massachusetts, 1962.

6. Dorfman, R.; "Formal Models in the Design of Water Resource Systems," Water
    Resources Research 1(3), 329-336 (1965).

7. Eastman, J., and C. Revelle; "Linear Decision Rule in Reservoir Management
    and Design, 3, Direct Capacity Determination and Intraseasonal Constraints,"
    Water Resources Research 9(1), 29-42 (1973).

8. Hufschmidt, M. M.; "Field Level Planning of Water Resource Systems," Water
    Resources Research 1(2), 147-163 (1965).

9. IBM; Mathematical Programming System/360, (360A-CO-14X) Linear and Separable
    Programming User's Manual H20-0476-1, 1968.

10. IBM; Mathematical Programming System Extended (MPSX), Linear and Separable
    Programming Program Description SH20-0968-0, 1971.

11. Jacoby, H. D., and D. P. Loucks; "Combined Use of Optimization and Simulation
    Models in River Basin Planning," Water Resources Research 8(6), 1401-1414
    (1972).

12. Loucks, D. P.; "Stochastic Methods for Analyzing River Basin Systems,"
    Technical Report No. 16, Water Resources and Marine Sciences Center, Cornell
    University, 1969.

13. Maass, A., and M. M. Hufschmidt; "Methods and Techniques of Analysis of the
    Multiunit, Multipurpose Water-Resource System: A General Statement," in
    Design of Water Resource Systems, as in 5.

14. Marks, D. H., J. C. Schaake, and D. C. Major; "Formulation and Evaluation of
    Alternatives in the Design of Water Resource Systems," as in 4.

15. Poblete, J-A., and R. T. McLaughlin; "Time Periods and Parameters in Mathe-
    matical Programming for River Basins," Technical Report No. 128, Ralph M.
    Parsons Laboratory for Water Resources and Hydrodynamics, Massachusetts
    Institute of Technology, 1970.

16. Revelle, C., E. Joeres, and W. Kirby; "Linear Decision Rule in Reservoir
    Management and Design, 1, Development of the Stochastic Model," Water
    Resources Research 5(4), 767-777 (1969).

17. Revelle, C., and W. Kirby; "Linear Decision Rule in Reservoir Management and
    Design, 2, Performance Optimization," Water Resources Research 6(4), 1033-
    1044 (1970).

18. Rogers, P.; "The Role of Optimizing Models in Resource Planning," presented
    at the Seminar on Development of Water Resources in Maracaibo, Venezuela
    (1968).

19. Rogers, P.; "A Game Theory Approach to the Problems of International River Basins," Water Resources Research 5(4), 749-760 (1969).

20. Smith, D. V.; "Stochastic Irrigation Planning Models," Technical Report, Center for Population Studies, Harvard University, 1970.

21. Wagner, H. M.; Principles of Operations Research, Prentice-Hall, Englewood Cliffs, New Jersey, 1969.

22. Wallace, J. R.; "Linear-Programming Analysis of River-Basin Development," Ph.D. Thesis, Dept. of Civil Engineering, Massachusetts Institute of Technology, 1966.

APPENDIX: *List of Variables*

Variable	Definition	Decision Variable ($\checkmark$ = yes, – = no)
X^1_{ij}	Reservoir head at the beginning of season j at site i.	$\checkmark$
X^2_{ij}	Average release from reservoir i during season j.	$\checkmark$
Y^1_{ij}	Average interbasin export at site i during season j.	$\checkmark$
Y^2_{ij}	Average diversion for irrigation at site i during season j.	$\checkmark$
Y_i	Maximum average export during the year from site i.	$\checkmark$
e	Power plant efficiency.	–
F_{ij}	Average streamflow at site i during season j.	–
z_i	Power plant capacity at site i.	$\checkmark$
h_j	Number of hours in season j.	–
W^1_{ij}	Volume of water supplied for irrigation at site i during season j.	$\checkmark$
W^2_{ij}	Average interbasin import at site i during season j.	$\checkmark$
W_i	Maximum average import during the year at site i.	$\checkmark$
Q_{ij}	The inflow at site i during season j.	$\checkmark$
k_j	Number of seconds in season j.	–
U^1_{ij}	Amount of land irrigated at site i during season j.	$\checkmark$
U^2_{ij}	Hydroelectric energy produced at site i during season j.	$\checkmark$
U_i	Amount of land irrigated during the whole year at site i.	$\checkmark$
V^1_{ij}	Average return flow from site i during season j.	$\checkmark$
V^2_{ij}	Volume of water stored at site i at the beginning of season j.	$\checkmark$
V_i	Maximum amount of water stored at site i.	$\checkmark$

Variable	Definition	Decision Variable ($\checkmark$ = yes, − = no)
ρ_{ij}	Power load factor at site i during season j.	−
α_i	Unit capital cost for reservoir i.	−
β_i^I	Unit gross benefit derived from agricultural output at site i.	−
β_{ij}^P	Unit gross benefit derived from energy production at site i during season j; equivalently, the market price for energy produced at site i during season j.	−
γ_i^E, γ_i^I	Unit capital costs for building interbasin transfer channels at site i.	−
δ_i	Unit capital cost for a power plant at site i.	−
ε_{ij}	Water loss coefficient for flow through irrigation diversion canals at site i during season j.	−
μ_{ij}	Total irrigation loss coefficient at site i during season j.	−
ν_{ij}	Unit gross benefit derived from interbasin export of water from site i during season j.	−
ρ_{ij}	Consumption coefficient for irrigation at site i during season j.	−
σ_i	Storage-head coefficient at site i.	−
τ_{ij}	Irrigation water-use coefficient at site i during season j.	−
ϕ_i	Unit capital cost of irrigation works at site i.	−

Chapter 13

A LINEAR PROGRAMMING ANALYSIS OF PAPER RECYCLING *

C. Roger Glassey
Department of Industrial Engineering
and Operations Research
University of California, Berkeley

and

Virendra K. Gupta
Operations Research Center
University of California, Berkeley

*This research has been partially supported by the National Science Foundation
under Grant GK-23153 and the Public Health Service Research Grant ECO0260-05 from
the Environmental Control Administration--Bureau of Solid Waste Management, U.S.
Department of Health, Education,and Welfare with the University of California.   It
is intended to provide some guides for material policy relative to recycling.

Abstract. A simple linear programming model is
constructed for production use, and recycling of
various paper and related products. This model
is used to estimate the maximum feasible recycling
rate given the current state of pulp and paper
technology. Parametric linear programming is used
to investigate the effect of the efficiency of
collecting, sorting, and transporting waste paper
on the fraction recycled.

1. *Introduction*

In recent years, interest in paper recycling has grown, as part of the increased popular awareness of environmental quality problems, for at least two
reasons. First, paper makes up about half of the weight of municipal refuse, and
municipal refuse disposal is an increasingly difficult problem. Landfill sites
near large cities, which offer the cheapest disposal method, are filling up at an
increasing rate as the per capita solid waste generation continues to increase.
Costs of acquiring more distant sites also continue to rise, as do transportation
costs; hence reducing the total amount of refuse for disposal is of increasing
economic importance aside from the environmental quality issue. Secondly, the
current rate of increase of wood pulp use (3.5% per year) extrapolated from the 54
million tons domestic consumption of 1970, approaches 100 million tons per year by
1988 with the resulting increased pressure on forest resources. During this
period the demand for competitive uses of these resources (for lumber and recreation) is likewise expected to increase.

In this study, a simple model is constructed for the production, use, and recovery of paper and related products made from wood fiber. Henceforth, the term
"paper" will refer to the aggregate of such products. This model will be used to
answer the following questions:

(1) What is being done at present - how much paper is being recycled and
from what sources?

(2) Given the present paper making technology, how much paper can be
easily recycled?

In considering these, we shall be dealing mainly with the technological constraints and shall ignore the economics of waste paper recovery primarily because
cost data are very difficult to obtain.

The starting point of the analysis is the simple flow model of Figure 1. In
this model the three major processes are paper production, consumption, and waste
paper recovery. Principal exogenous flows are paper imports and exports, permanent stock increases (books in libraries, building paper in new construction,

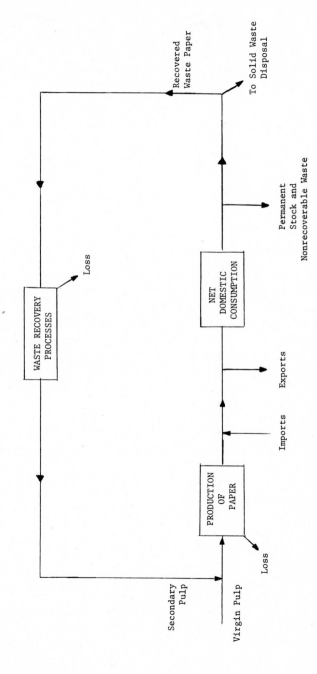

FIGURE 1. PAPER PRODUCTION, USES, RECYCLING: A FLOW MODEL

etc.), nonrecoverable waste (paper burned in fireplaces, flushed down the sewage system, etc.), and the paper fraction of the solid waste currently entering land fills, incenerators, and other disposal methods. The principal source of fiber for paper is virgin wood pulp supplemented by recovered "secondary pulp."

If imports, exports, and net domestic consumption remain fixed (at the 1970 levels) it is clear that minimizing virgin pulp input to the system is equivalent to minimizing the waste paper output to solid waste disposal, which is equivalent to maximizing secondary pulp use.

Certain technological constraints in the papermaking process prevent complete utilization of all waste paper (corresponding to zero flow to waste disposal) and these constraints are specific to various types of paper. Thus, the flows in Figure 1 must be regarded as vector valued, and a more detailed examination of the papermaking industry is required.

2. The Problem

Types of Paper Production

New paper is broadly classified into 13 categories (Ref. 2) which are shown in Table 1 and are also listed below.

1. Newsprint
2. Uncoated groundwood paper
3. Coated paper
4. Uncoated book paper
5. Writing and related paper
6. Bleached bristols
7. Unbleached kraft packaging and industrial covering paper
8. Other packaging and industrial covering paper
9. Tissue paper
10. Unbleached kraft linerboard
11. Bleached packaging
12. Box board
13. Building paper, board, etc.

Types of Waste Paper and Sources

These papers, after use, go to various types of waste. The Paper Stock Institute of America has established 46 standard types of waste paper (Ref. 10). These are classified into 11 main groups as listed below.

No.	Description	Produced (p_i)	Imported	Exported	Consumed (c_i)
	Paper				
1.	Newsprint	3,309	6,635	114	9,800
2.	Uncoated groundwood paper	1,165	234	7	1,392
3.	Coated paper	3,237	4	77	3,164
4.	Uncoated book paper	2,572	72.5	11	2,633.5
5.	Writing and related paper	2,811	27	114	2,724
6.	Bleached bristols	1,061	–	–	1,061
7.	Unbleached kraft packaging and industrial covering paper	3,723	35	156	3,602
8.	Other packaging and industrial covering paper	1,595	31	73	1,553
9.	Tissue paper	3,671	6.5	33	3,644.5
	Paperboard				
10.	Unbleached kraft linerboard and medium corrugated	11,518	20	1,766	9,732
11.	Bleached packaging	7,036	–	129	6,907
12.	Box board	6,875	2	273	6,603
13.	Building paper, board and construction paper	3,812	201	59	3,954
	Total paper and paperboard	52,384	7,268	2,842	56,810
	Waste Paper		114	408	

TABLE 1.   PRODUCTION, IMPORT, EXPORT, AND CONSUMPTION OF PAPER AND PAPERBOARD
            IN 1970 (from Ref. 1 and 2, in thousands of tons)

	Description	Types of Waste Paper as Listed in Ref. 10
1.	News	6-8, 24, 25
2.	Groundwood shavings	22, 23, 26
3.	Flyleaf shavings	27
4.	Magazines and books	28, 29, 42, 43
5.	Waste ledger	30, 38-41
6.	Envelopes and tab cards	31-37, 44-46
7.	Kraft bags and cuttings	14-17, 19-21
8.	Corrugated cuttings	9-13
9.	New colored krafts	18
10.	Box board cuttings	4
11.	Mixed waste paper	1-46

Table 2 shows the types of paper which go into a particular type of waste paper. For example, column 1 in Table 2 means that types of paper 1 (newsprint) and 2 (uncoated groundwood paper) compose waste paper of type 1 (news). Also, not all types of waste paper can be used for making each type of new paper. The array in Table 3 shows the type of waste paper which can be used for making each type of new paper.

| | Type of Waste Paper | | | | | | | | | | |
Type of Paper	1	2	3	4	5	6	7	8	9	10	11
1	1	0	0	0	0	0	0	0	0	0	1
2	1	1	0	0	0	0	0	0	0	0	1
3	0	1	1	0	0	0	0	0	0	0	1
4	0	1	1	1	0	0	0	0	0	0	1
5	0	0	0	0	1	0	0	0	0	0	1
6	0	0	0	0	0	1	0	0	0	0	1
7	0	0	0	0	0	0	1	0	0	0	1
8	0	0	0	0	0	0	1	0	0	0	1
9	0	0	0	0	0	0	0	0	0	0	1
10	0	0	0	0	0	0	0	1	0	0	1
11	0	0	0	0	0	0	0	0	1	1	1
12	0	0	0	0	0	0	0	0	0	1	1
13	0	0	0	0	0	0	0	0	0	0	1

TABLE 2.   POSSIBLE WASTE PAPER COMPOSITION OF NEW PAPER
                  (Results based on a field study.)

Minimum Virgin Pulp Requirements

Table 4 shows estimates of the minimum fraction of virgin pulp required to maintain the required quality level of each of the 13 types of paper. Of course, the virgin pulp fraction can be (and usually is) higher than this minimum. These estimates are based on extensive conversations with representatives of several paper manufacturers*. The upper estimate is a fraction that is the opinion of the experts. It will usually yield acceptable quality. The lower estimate might also be satisfactory. It should be noted that these estimates are not the results of extensive experimentation. The data of Table 3 is likewise based on the same expert opinions. However, the conclusions are solely those of the authors, not of the experts.

---

*A list of those contacted is in Appendix I. We are very grateful for their assistance.

Type of Paper	1	2	3	4	5	6	7	8	9	10	11	De-inking	Asphalt Dispersion	Bleaching	Fiber Loss (%)
1	1	0	0	0	0	0	0	0	0	0	0	x			15***
2	1	1	0	1	1	0	0	0	0	0	0	x		x	10
3	0	1	1	1	1	1	0	0	0	0	0	x		x	15
4	0	1	1	1	1	1	0	0	0	0	0	x		x	15
5	0	0	0	0	1	1	0	0	0	0	0	x		x	10
6	0	0	0	1	1	1	0	0	0	0	0	x		x	5
7	0	0	0	0	0	0	1	1	0	0	0				20
8	0	0	0	0	0	0	1	1	0	0	0				20
9	1	0	0	1	0	0	0	0	0	0	0	x			15
10	0	0	0	0	0	0	0	1	0	0	0		x		10
11	0	0	0	0	0	1	0	0	0	0	0			x	5
12	1	1	1	1	1	1	1	1	1	1	1	x**		x**	7
13	1	0	0	0	0	1	0	1	0	0	1	x**		x**	10

* Results based on a field study (see Assumption D, Section 6).

** These are sometimes used according to the particular type of paper to be made.

*** The various details for newsprint are from the Garden State Paper Co., Los Angeles, California.

TABLE 3.   TYPE OF WASTE PAPER USED IN MANUFACTURING PAPER AND FIBER LOSS*

Description of the Processes

There are three main processes which are used for converting waste paper into re-usable secondary pulp:

(i) De-inking

In order to produce white pulp from waste papers suitable for book papers, it is necessary to remove the ink from the waste paper. The process of doing this is known as de-inking. It consists of sorting, dusting, defibering, cooking, washing, bleaching, refining, and screening of waste paper. There are two basic steps in de-inking.

(a) Dissolving or loosening the ink and defibering the stock by chemical means.

(b) Removing the ink from the pulp by mechanical washing.

No.	Description	Lower Estimate		Upper Estimate	
		%	Qty. (10^3 tons)	%	Qty. (10^3 tons)
	Paper				
1.	Newsprint	0	0	0	0
2.	Uncoated groundwood paper	50	583	60	699
3.	Coated paper	40	862	50	1,077
4.	Uncoated book paper	40	1,082	50	1,285
5.	Writing and related paper	60	1,685	70	1,965
6.	Bleached bristols	70	742	80	848
7.	Unbleached kraft packaging and industrial covering paper	80	2,980	80	2,980
8.	Other packaging and industrial covering paper	80	1,275	80	1,275
9.	Tissue paper	10	367	10	367
	Paperboard				
10.	Unbleached kraft linerboard	80	9,210	80	9,210
11.	Bleached packaging	90	6,320	90	6,320
12.	Box board	0	0	0	0
13.	Building paper, board and construction paper, etc.	0	0	10	381
	TOTAL OF VIRGIN PULP		25,052		26,407
	SECONDARY PULP		28,803		27,448
	TOTAL PULP REQUIRED		53,855		53,855

TABLE 4.   THE ESTIMATED MINIMUM REQUIREMENTS FOR VIRGIN PULP (1970 production, 5% process loss assumed)

De-inking has been described basically as a laundary problem. Wetting, detergent action, and dispersion are involved in the suspension and removal of ink, sizing, coating material and fillers which are present in the waste papers.

(ii) Asphalt Dispersion

In certain packing and covering paper, asphalt coating is provided to protect it from the environment. This asphalt creates a problem in cylinder machine mills which use waste paper for linerboard stock. In recent years this difficulty has been overcome by the development of an asphalt dispersion system, which is used after repulping and cleaning, and before refining. There are three basic steps in the dispersing

process, namely, melting the asphalt, mechanically dispersing the
asphalt, and chilling and fixing the asphalt.

(iii) <u>Bleaching</u>

The waste paper after de-inking or after the asphalt dispersion
process is sometimes bleached depending upon its use in a particular
type of new paper to be made. The objective of bleaching is the production of a white pulp of stable color. The selection of the process
depends on the type of waste paper to be used and the type of new
paper to be made. Various details regarding the processes used for
making different varieties of papers are shown in Table 3.

Process Losses

There are two main types of losses which usually occur as shown in the flow
model appearing in Figure 1. These are:

(i) The loss in manufacturing paper from pulp. This is the loss of very
fine fiber which passes through the screens and cannot be reclaimed.
It is estimated to be approximately 5% for all types of papers.

(ii) The loss in waste recovery processes. The loss in these processes
depends on the type of waste paper used and are different for different
types of paper (see Table 3).

Production, Imports, Exports, and Net Domestic Consumption

Figures for the year 1970 are shown in Table 1. Import and export of waste
paper are also shown in Table 1, but is such an insignificant fraction of the
total annual production that it is ignored.

Inventory Accumulation and Nonrecoverable Waste

As estimated by the American Paper Institute (Ref. 3), over 7 million tons of
paper remained in permanent use in such forms as books and building materials or
were disposed of in home fireplaces or, in the case of such products as tissues,
in sewer systems. Unfortunately, data on the specific type of paper thus nonrecoverable is not available. Some guesses are made and are shown in Table 5 consistent with a total of about 7 million tons. The reader may form his own
opinions about the plausibility of these guesses.

No.	Domestic Consumption (c_i)	Paper Going to Permanent Stock or Nonrecoverable Use		Paper Going to Waste Channels (s_i)
		%	wt. (ℓ_i)	
1.	9,800	0	0	9,800
2.	1,392	0	0	1,392
3.	3,164	10	316.0	2,848
4.	2,633.5	40	1,051.5	1,582
5.	2,724	20	545.0	2,179
6.	1,061	0	0	1,061
7.	3,602	0	0	3,602
8.	1,553	0	0	1,553
9.	3,644.5	90	3,279.5	365
10.	9,732	0	0	9,732
11.	6,907	0	0	6,907
12.	6,603	0	0	6,603
13.	3,954	60	2,370.0	1,584
TOTAL	56,810		7,562*	49,248

* The estimated figure is approximately equal to the available data on the total quantity going to permanent use or destroyed (Ref. 3).

TABLE 5. PAPER GOING TO VARIOUS CHANNELS AFTER USE (in 10^3 tons, 1970 data)

3. The Model

Assumptions

A. Internal recycling of mill scrap is ignored in the model. It is assumed that such scrap is reused inside the mills and does not go into the post-consumer waste channels of the model.

B. A linear relation between inputs and outputs hold for all processes.

C. If different kinds of waste papers are used for making a certain type of paper, they could be mixed in any ratio.

D. In the subsequent analysis, all data and computations are made on the basis of dry weight tons. We assume that, within the system diagrammed in Figure 1, fiber is conserved, thus all increments and losses of fiber are explicit in the exogeneous flow. For most types of paper, this is approximately equivalent to a conservation of mass since the dry weight of the paper products is essentially the weight of the fiber. The notable exception to this rule is the coated papers, in which the clay coating is 30% to 40% of the weight of the finished product. Hence, the output of the production process is considerably larger than the input weight of pulp, and a corresponding loss in weight appears in the process of recovering fiber from used coated paper.

Variables

X_{1i} The number of tons of virgin pulp to be used for paper type i.

X_{2i} The number of tons of secondary pulp to be used for paper of type i.

Y_{ij} The number of tons of paper type i converted into waste paper type j.

Z_{ji} The number of tons of waste paper j used to produce paper type i.

Constraints

The total fiber input to each production process, from both virgin and secondary pulp, must equal the output plus the 5% process loss. Since the tonnage output of coated papers includes the weight of the clay coating (about 30%), the appropriate input/output ratio is .737. Thus we have constraint (1).

$$X_{1i} + X_{2i} = 1.053p_i \quad \text{for} \quad i = 1,2,4,\ldots,13$$
$$= .737p_i \quad \text{for} \quad i = 3, \tag{1}$$

where p_i is the required production of product i.

Inequality (2) guarantees that the amount of virgin pulp in paper type i is at least as large as the minimum required. Or

$$X_{1i} \geq k_i p_i \quad\quad\quad \text{for} \quad i = 1,\ldots,13, \tag{2}$$

where k_i is the minimum fraction of virgin pulp in the production of product i (from Table 5).

Constraint (3) expresses the limitation on the amount of secondary fiber available for product i as recovered from the various grades of waste paper. The coefficient a_{ji} is the yield (one minus the fiber loss from Table 3) of the recovery process that converts waste grade j to paper type i. If waste grade j is not allowed in paper type i (as indicated by a zero in the column j, row i entry of Table 3) then a_{ji} is zero. In this case, the variable Z_{ji} may be omitted from the linear program.

$$X_{2i} \leq \sum_{j=1}^{11} a_{ji} Z_{ji} \qquad \text{for} \quad i = 1,\ldots,13. \qquad (3)$$

Inequality (4) expresses the limit on the total amount of waste of grade j available from recovered paper of type i. The coefficients b_{ij} are shown in Table 2 and equal 1 if paper type i is an allowable constituent of waste type j, and zero otherwise.

$$\sum_{i=1}^{13} Z_{ji} \leq \sum_{i=1}^{13} b_{ij} Y_{ij} \qquad \text{for} \quad j = 1,\ldots,11. \qquad (4)$$

The material balance equations for the recovered waste S_i, where Y_{io} is the tonnage not utilized in any recovery process are in (5).

$$\sum_{j=1}^{11} b_{ij} Y_{ij} + Y_{io} = S_i \qquad \text{for} \quad i = 1,\ldots,13. \qquad (5)$$

Since the estimates of the fraction of each type of paper flowing into permanent storage and the nonrecoverable losses is subject to considerable uncertainty, the sensitivity of the model to the fraction of used paper available for recovery was investigated by means of parametric linear programming. The parameter λ, ranging from 0 to 1, represents the fraction of potentially available waste that is recovered, and is the same for all types of paper. Thus, from domestic consuption c_i, we subtract the estimated nonrecoverable waste and inventory increment ℓ_i, and then scale the difference which is the potentially available waste, by the factor λ. To be precise, we set

$$S_i = \lambda(c_i - \ell_i) \qquad \text{for} \quad i = 1,2,4,\ldots,13$$

$$S_i = .7\lambda(c_i - \ell_i) \qquad \text{for} \quad i = 3,$$

where c_i is the domestic consumption of paper of type i in tons (Table 5), ℓ_i is the addition to inventory plus nonrecoverable losses in tons (Table 5), and λ is the fraction of total available paper going to recoverable waste paper after use.

The objective function selected is the total virgin pulp used, and the minimum value of this quantity is obtained as a function of λ, the recovery fraction (see Fig. 6). Thus, the linear programming model is to minimize $V(\lambda) = \sum_{i=1}^{13} X_{1i}$, subject to constraints (1) through (5), and the nonnegativity requirements on all variables. The parametric linear programming algorithm first solves the linear program when λ is fixed (at zero) and the right hand side of (5) is thus fixed. The value of λ is then increased until the current basis is infeasible. A single exchange of basic variables is then sufficient to determine a new feasible (and optimal) solution. The λ value is increased again, and the process repeats until the maximum value of $\lambda = 1$ is reached. By doing this the linear program is solved for all values of λ between 0 and 1 inclusively.

4. Results and Conclusions

When the constraints (1) through (5) are simplified to omit impossible activities (e.g., leaving out the variable Y_{ij} if the coefficient b_{ij} is zero), the resulting linear program has 63 constraints (rows) and 96 variables (columns). Two linear programs were set up and solved; one using the lower estimates of Table 4 as the k_i values in constraints (2); the other using the upper estimates. For each of these programs, the parametric analysis was performed with λ, the recovery fractions, varying between 0 and 1. The graph of $V(\lambda)$ corresponding to the upper estimate of virgin pulp requirements for each product is shown in Figure 6, and the details of the two solutions for $\lambda = 1$ are presented in Figures 2, 3, 4 and 5.

From these results, the following general conclusions can be drawn:

(i) With no change in papermaking technology, it is possible to reduce the annual virgin pulp consumption from the 1970 level of 45 million tons (or 83% of the total pulp requirement) to 28 million tons or 52% of the total requirement, assuming that 70% of the potentially recoverable paper is recycled. If the lower estimates of virgin pulp requirement are used, this figure is 27.5 million tons or 51% at the 70% recovery level. The reduction in the national solid waste disposal problem would be of the order of 17 million tons annually. Assuming the average cost for collection and disposal is $14 per ton (Ref. 4), the potential saving is on the order of $238 million annually, which presumably could be used to offset the costs of collecting and processing waste paper.

(ii) Attempting to reduce the flow of paper into solid waste disposal systems below 30% of the potentially salvageable paper (i.e., increasing λ above .7) results in very small improvement, as seen by the rapid decrease in slope of the $V(\lambda)$ curve. The binding constraints that keep virgin pulp consumption above about 26 million tons are primarily the minimum virgin pulp requirements of Table 4, not the restrictions on the flow of fiber from one type of paper to another as shown in the information matrices of Tables 2 and 3.

This conclusion can be verified by comparing the optimal linear programming solutions for the upper and lower pulp requirement estimates (at $\lambda = 1$) with the total minimum virgin pulp requirements shown in Table 4. These are 25.052 and 26.407 million tons respectively; the calculations producing them ignores the limitations on sources of secondary pulp. The linear programming solutions, which do consider these restrictions, have objective function values of 25.1 and 26.5

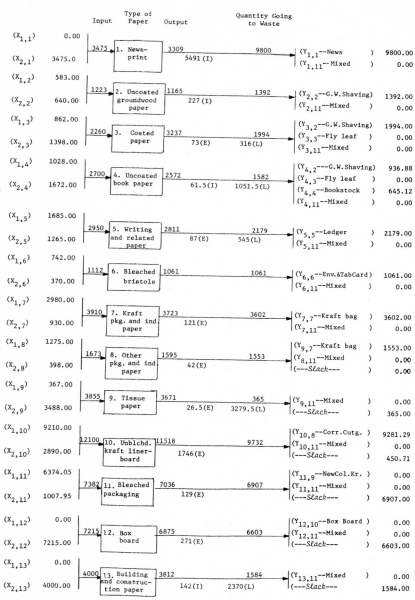

L ≡ Lost or in permanent stock, I ≡ Import, E ≡ Export

FIGURE 2.   L.P. SOLUTION - LOWER ESTIMATE

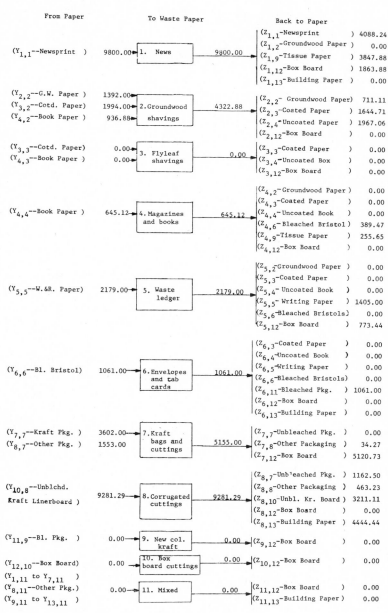

FIGURE 3.   LINEAR PROGRAMMING SOLUTION (with λ=1) - LOWER ESTIMATE

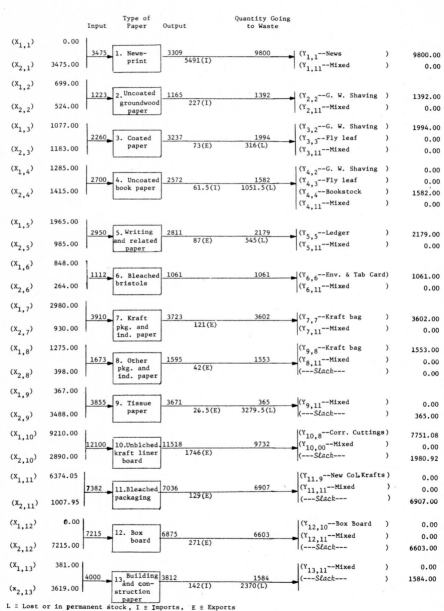

		Input	Type of Paper	Output	Quantity Going to Waste		
$(X_{1,1})$	0.00						
		3475	1. News-print	3309 5491(I)	9800	$(Y_{1,1}$--News)	9800.00
$(X_{2,1})$	3475.00					$(Y_{1,11}$--Mixed)	0.00
$(X_{1,2})$	699.00						
		1223	2. Uncoated groundwood paper	1165 227(I)	1392	$(Y_{2,2}$--G. W. Shaving)	1392.00
$(X_{2,2})$	524.00					$(Y_{2,11}$--Mixed)	0.00
$(X_{1,3})$	1077.00					$(Y_{3,2}$--G. W. Shaving)	1994.00
		2260	3. Coated paper	3237 73(E)	1994 316(L)	$(Y_{3,3}$--Fly leaf)	0.00
$(X_{2,3})$	1183.00					$(Y_{3,11}$--Mixed)	0.00
$(X_{1,4})$	1285.00					$(Y_{4,2}$--G. W. Shaving)	0.00
		2700	4. Uncoated book paper	2572 61.5(I)	1582 1051.5(L)	$(Y_{4,3}$--Fly leaf)	0.00
$(X_{2,4})$	1415.00					$(Y_{4,4}$--Bookstock)	1582.00
						$(Y_{4,11}$--Mixed)	0.00
$(X_{1,5})$	1965.00						
		2950	5. Writing and related paper	2811 87(E)	2179 545(L)	$(Y_{5,5}$--Ledger)	2179.00
$(X_{2,5})$	985.00					$(Y_{5,11}$--Mixed)	0.00
$(X_{1,6})$	848.00						
		1112	6. Bleached bristols	1061	1061	$(Y_{6,6}$--Env. & Tab Card)	1061.00
$(X_{2,6})$	264.00					$(Y_{6,11}$--Mixed)	0.00
$(X_{1,7})$	2980.00						
		3910	7. Kraft pkg. and ind. paper	3723 121(E)	3602	$(Y_{7,7}$--Kraft bag)	3602.00
$(X_{2,7})$	930.00					$(Y_{7,11}$--Mixed)	0.00
$(X_{1,8})$	1275.00					$(Y_{9,8}$--Kraft bag)	1553.00
		1673	8. Other pkg. and ind. paper	1595 42(E)	1553	$(Y_{8,11}$--Mixed)	0.00
$(X_{2,8})$	398.00					(---Slack---)	0.00
$(X_{1,9})$	367.00						
		3855	9. Tissue paper	3671 26.5(E)	365 3279.5(L)	$(Y_{9,11}$--Mixed)	0.00
$(X_{2,9})$	3488.00					(---Slack---)	365.00
$(X_{1,10})$	9210.00					$(Y_{10,8}$--Corr. Cuttings)	7751.08
		12100	10. Unblched. kraft liner board	11518 1746(E)	9732	$(Y_{10,00}$--Mixed)	0.00
$(X_{2,10})$	2890.00					(---Slack---)	1980.92
$(X_{1,11})$	6374.05					$(Y_{11,9}$--New Col.Krafts)	0.00
		7382	11. Bleached packaging	7036 129(E)	6907	$(Y_{11,11}$--Mixed)	0.00
$(X_{2,11})$	1007.95					(---Slack---)	6907.00
$(X_{1,12})$	0.00					$(Y_{12,10}$--Box Board)	0.00
		7215	12. Box board	6875 271(E)	6603	$(Y_{12,11}$--Mixed)	0.00
$(X_{2,12})$	7215.00					(---Slack---)	6603.00
$(X_{1,13})$	381.00					$(Y_{13,11}$--Mixed)	0.00
		4000	13. Building and construction paper	3812 142(I)	1584 2370(L)	(---Slack---)	1584.00
$(X_{2,13})$	3619.00						

L ≡ Lost or in permanent stock, I ≡ Imports, E ≡ Exports

FIGURE 4. LINEAR PROGRAMMING (with λ=1) - UPPER ESTIMATE

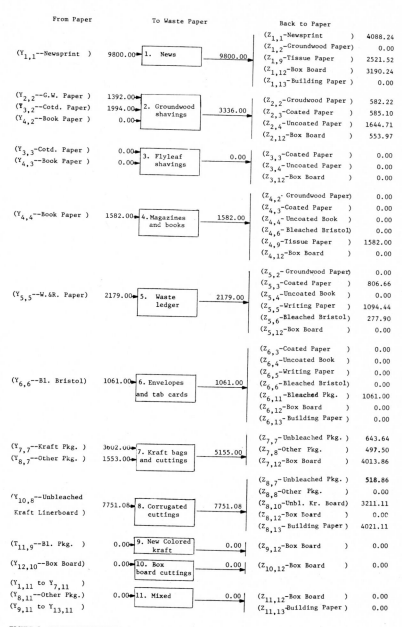

FIGURE 5. LINEAR PROGRAMMING (with λ=1) - UPPER ESTIMATE

million tons, corresponding to the same lower and upper estimates. The conclusion
that the minimum virgin pulp requirements are the binding constraints is rein-
forced by examining in detail the values of the dual variables corresponding to
the virgin pulp requirements for the 13 kinds of paper. These dual variables may
be loosely interpreted as the rate of change of the optimal value of the objective
function with respect to the change in the right hand side of the constraints. In
the optimal solution corresponding to the lower estimate, 9 of these 13 dual variables have value 1.0; in the upper estimate solution 10 have value 1.0. Thus, a
reduction of 1 ton in the minimum virgin pulp required for any of these products
results in a reduction of the objective value by 1 ton.

(iii) Extrapolating the $V(\lambda)$ curve at its initial slope from $\lambda = 0$ to $\lambda = 1$,
along the line A - B' in Figure 6, gives an estimate of about 9.5 million tons per
year as the rock bottom minimum virgin pulp requirement to make up process losses
and inventory increment. This figure assumes that every scrap of used paper is
recycled.

Since no economic factors were considered in this model, it is not suitable
for predicting what level of recycling might be attained. It does, however, provide an estimate of what is feasible with present technology.

References

1. American Paper Institute; "Monthly Statistical Summary," February and March
    1971.

2. American Paper Institute; "A Test of Stamina," New York, 1970.

3. American Paper Institute; "The Paper Industry's Part in Protecting the
    Environment," New York, 1970.

4. California Waste Management Study, Aerojet General Corporation, Report No.
    3056 (Final), August 1965.

5. Carr, W. F.; "Increased Wood Fiber Recycling: A Must," Forest Product
    Laboratory Report, Wisconsin, August 1970.

6. Crysler, F. S., "Solid Waste Management: Paper and Board in Perspective,"
    presented at the spring meeting, Paperboard Group, American Paper Institute,
    May 1971.

7. George, P. C.; "The CMI Report on Solid Waste Control," Washington, D. C.,
    1970.

8. Gupta, V. K.; "Supplementary Report on Paper Recycling," Operations Research
    Center Report, the University of California at Berkeley, 1971.

9. Institute of Solid Wastes of American Public Works Association, "Municipal
    Refuse Disposal," 1970.

10. Paper Stock Institute of America, "Paper Stock Standards and Practices -
    Circular PS-70," New York, January 1, 1970.

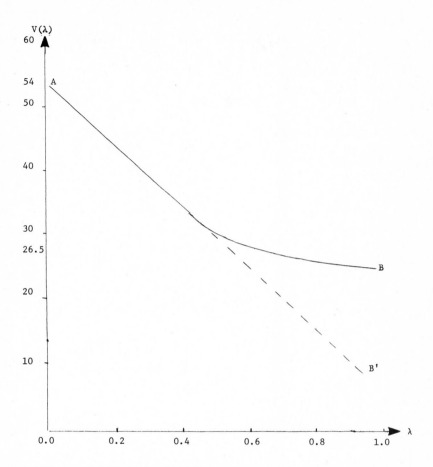

FIGURE 6.   FRACTION OF WASTE RECOVERABLE VERSUS THE MINIMUM VALUE OF VIRGIN
           PULP (in MILLION TONS)

APPENDIX: *List of Companies Contacted*

	Name of Company	Person(s) Contacted
1)	Crown Zellerbach Corporation 1 Bush Street San Francisco, California 94104	Mr. R. A. Butler Vice President Manufacturing
	Crown Zellerbach Corporation 1 Bush Street San Francisco, California 94104	Mr. Lowell M. Clucas Director of Corporation Communications
2)	Fibreboard Corporation P. O. Box 3611 San Francisco, California 94119	Mr. Thomas F. Martin Manufacturing Manager Paperboard Division
3)	Consolidated Fibres 1800 Peralta Oakland, California 94607	Mr. George Graham
4)	Kaiser Gypsum Company, Inc. P. O. Box 2018 San Leandro, California 94577	Mr. Chuck B. Langenstein Plant Engineer
5)	Garden State Paper Co., Inc. 2205 West Mt. Vernon Avenue Pamona, California 91766	Mr. Emerson Lewis Mill Manager
6)	Black Clawson Company Middletown, Ohio 45042	Mr. W. Herbert Director of Engineering Shartle/Pandia Division
7)	Pulp & Paper 500 Howard Street San Francisco, California 94105	Mr. W. C. Williams Editorial Director
8)	Western Kraft Corporation Richmond, California	Mr. James Quick
9)	Independent Paper Stock Company Emeryville, California	

Chapter 14

MODELLING THE DISTRIBUTION OF INTERREGIONAL FREIGHT TRAFFIC
IN GREAT BRITAIN*

Pat O'Sullivan
Department of Geography/Transportation Center
Northwestern University
Evanston, Illinois

*The work reported here was part of a Social Science Research Council (U.K.)
project, jointly directed by M. D. I. Chisholm and the author. Many thanks are
due to John Adams, Bob Dennis, and Mike Dunford for their computational assistance. A variation of this article originally appeared in the *Journal of
Regional and Urban Economics* 1(4), 383-396 (1972).

Abstract. This chapter illustrates the efficacy
of linear programming (LP) as a method of predicting flows of commodities between regions of
Great Britain. Transportation problem solutions
were found for the annual movement by road of
eleven commodity categories between 78 origin and
destination zones in 1964. Road distances were
used in place of freight costs. Data on actual
flows of these commodities in that year were available for comparison. For the more homogeneous
commodity classes the LP solution replicated the
actual pattern quite well.

Another experiment was carried out, using the
same commodities, on a subset of zones which were
more similar in size--namely, the 24 major cities
in Great Britain; the correspondence between LP
solutions and actual flows was much closer than
for all 78 zones.

A further test involved finding LP solutions
for total road plus rail traffic with six mineral
and metal commodities which use the rail system
extensively. These solutions for total flows
proved to correlate with actual flows better than
solutions for road movement alone.

The "shadow prices" of the minimizing dual of
the transportation problem were produced and
examined to see if they made sense in terms of
geographical comparative cost advantage variations
for production and consumption.

1. *The Problem*

This chapter reports some first tentative attempts at modelling the demand
for interregional freight transport in Great Britain, with a view to developing
a forecasting capability. The desire to forecast demand for this traffic is
motivated by several objectives. Those concerned with investment in transport
systems obviously require such projections. Furthermore, a method of predicting
freight flows as a function of the location of economic activities and the costs
of movement between regions can be used to evaluate different proposals for
regional development in transport cost terms. Given alternative proposals for
the regional disposition of production and consumption, the model could be used
to estimate the demand for transport which each proposal generated. These demands
could be translated into transport costs and, thus, provide a means of comparing
the alternatives. In addition, a freight forecasting model would be of value in
evaluating the impact of transport facility improvements or regulatory changes on
the spatial structure of the economy.

There are several inviting theoretical and operational forms upon which to
base such a forecasting endeavor; however, these are mostly very demanding in
terms of data. We constrained ourselves to work within the scope of existing
empirical information. The limitations of the data available forced us to resort
to a staged approach to the task of modelling demand, in much the same manner as
that used in urban land use/transport planning. We first estimated generation and
attraction of traffic by various regions as a function of activity levels using
regression analysis and then examined different methods of linking these availabilities and requirements for various goods with a "distribution" model. The results of the traffic generation analysis are detailed in Chapter 5 of Chisholm and
O'Sullivan (Ref. 2). Tonnages of commodities generated and attracted by zones
were estimated as linear functions of employment in various industries and population in those zones using least squares--employment being a proxy for production
and population for consumption. Regional policy proposals are, of course, most
likely to be couched in terms of just such population-related variables as the
proxies we have chosen here, rather than in terms of production and consumption.

Assuming that satisfactory estimates of the tonnages of commodities available
and required by zones could be made for any proposed or projected regional allotment of activities, a method of distributing flows between origins and destinations was sought. Due to the coarseness of the data commodity classification,
our attention was first drawn to the "gravity model" formulation, especially the
form proposed by Wilson (Ref. 9). He derived this model as the expression for the
"most probable" configuration of flows which satisfies constraints on availabilities and requirements for the good in various zones and an "energy conservation"
constraint on total transport expenditures. The model predicts movements between
origins and destination (X_{ij}) in terms of the product of the total generation of
the origin (a_i), the total traffic attracted to the destination (b_j), and an inverse power function of the distance between the origin and destination $(d_{ij}^{-\beta})$,
modified by two factors (A_i, B_j) representing the solutions of the Lagrangeans
associated with the availability and requirements constraints (i.e., $\sum_j X_{ij} = a_i$ and
$\sum_i X_{ij} = b_j$); in particular

$$X_{ij} = A_i B_j a_i b_j d_{ij}^{-\beta},$$

where $A_i = (\sum_j B_j b_j d_{ij}^{-\beta})^{-1}$ and $B_j = (\sum_i A_i a_i d_{ij}^{-\beta})^{-1}$. These two sets of values
are solved for iteratively. The Lagrange multiplier (β) associated with the
transport expenditure constraint, which requires that total tonnage times distance
be a constant (i.e., $\sum_{ij} X_{ij} d_{ij} = C$), is not solved for but empirically determined
from an examination of current flow patterns.

Experiments with this model (Ref. 2, Chapter 6), suggested that the "most probable distribution" of the gravity model gave too even a distribution of flows to represent flows of commodities. It was felt, that despite the crudity of the commodity categories for which data were available, the actual flows might not be too different from the optimal as given by a linear programming solution. Thus simple transportation problem solutions might provide better estimates of future flows even for broad categories of goods. It is this conjecture which is tested empirically here.

2. *The Model*

A general description of transportation problems and their solution is given in Reference 3 and an economic interpretation of the dual values of the solution is discussed in Reference 8.

In the circumstances we are addressing, the linear programming transportation problem represents the optimal configuration of inter-regional trade in the primal and the equilibrium set of prices in origin and destination regions in its dual. Under the assumptions of a perfectly competitive market, fixed quantities available (a_i) and required (b_j) in various regions and known transport cost (d_{ij}) which vary linearly with volume traded (i.e., no scale economies in transport), transportation problem algorithms find the set of nonnegative flows (X_{ij}) between origin and destination regions which fulfill requirements at destinations and respect availabilities at origins while simultaneously minimizing the total cost of transport. The dual establishes the set of market clearing "shadow prices," which in equilibrium corresponds to the transport cost minimizing allocation of interregional trade.

Using the notation applied to the gravity model, with distance (d_{ij}) as a proxy for transport cost, and making the simplifying assumption that there is no trade surplus, i.e., $\sum_i a_i = \sum_j b_j$, we can write the primal problem as

minimize $\qquad C = \sum_{ij} d_{ij} X_{ij}$

subject to $\qquad \sum_j X_{ij} = a_i \quad$ for each origin i,

$\qquad\qquad\qquad \sum_i X_{ij} = b_j \quad$ for each destination j,

and $\qquad\qquad X_{ij} \geqq 0 \quad$ for all i, j.

The dual to this problem is

$$\text{maximize} \qquad V = \sum_j b_j V_j - \sum_i a_i U_i$$

$$\text{subject to} \qquad V_j - U_i \leq d_{ij} \qquad \text{for all } i,j.$$

It can be shown using certain economic interpretations (see Ref. 8) that the difference between the values of the variables U_i and U_k is a measure of the locational advantage of source i over source k. Similarly the difference $V_j - V_r$ can be interpreted as the locational advantage of destination j over destination r.

3. *Implementation*

In 1962, the Ministry of Transport (Ref. 5) conducted a road freight traffic survey which recorded flows of various commodities between the 107 zones into which they had divided Great Britain. To make these flows compatible with the results of a British rail survey of freight flows made in 1964, the commodities were consolidated to 13 groups and growth factors were applied to the flows which reflected national changes in production of the commodities between 1962 and 1964. The 107 regions were aggregated to 78. The resulting thirteen 78 by 78 road annual tonnage origin/destination matrices were made available by the Ministry (Ref. 6). The 13 commodity groups were: 1) coal and coke, 2) iron ore, 3) limestone, 4) scrap, 5) steel, 6) oil, 7) transport vehicles and equipment, 8) foodstuffs, 9) chemicals, 10) building minerals and materials, 11) other manufacturers, 12) other crude materials, and 13) miscellaneous. In fact, two of the road matrices, 2 and 3, were empty, so that there were only eleven commodity groups. The thirteen rail origin/destination matrices, and thus total road plus rail flows were also made available.

The row and column totals of these matrices give sets of availabilities and requirements (a_i and b_j) while the actual flows provided a basis for judging how well any model estimating flows replicated the real world. The totals of the row and column sums of these matrices do in fact equal, i.e., $\sum_i a_i = \sum_j b_j$.

The Ministry had also constructed a matrix of road distances between the centers of the 78 regions. The main diagonal of this matrix--the average length of journey within a zone--was estimated as 2/5 of the square root of the area of the zone. This is the appropriate moment of an area to represent average internal journey length, if it is assumed that the area is circular. Since transport costs vary monotonically with distance, at least, it seemed reasonable to use this distance matrix as a surrogate for interregional cost of transport.

Transportation problems were solved for movements by road of the eleven commodities between 78 origins and destinations.

To get a more comprehensive picture for the more homogeneous commodities with large volumes carried on the railways--including those exclusively carried by rail--solutions were found for total flows between 78 regions for the first six commodities. Road distances were used in these calculations as they were available and much effort would have had to be expended to extract rail distances and decide on the appropriate combinations of the two.

The 78 zones vary widely in area and industrial characteristics, which obviously distorts the picture presented by the data. To see if things improve if a more homogeneous subset of zones is used, solutions were found for flows of eleven commodities between the 24 major urban areas of Great Britain. This exercise was carried out by Dunford (Ref. 4).

Actual flows and LP solutions were compared by regression. The regression of LP solutions on actual flows suggested the manner and magnitude of the departure of one from the other. Perfect correspondence would result in an intercept value of zero and a slope of unity. As a further comparison, the number of occupied cells in the actual origin/destination matrices were counted to match against the maximum of m+n-1 positive values that are obtained from LP solutions. These were expressed as percentages of the total number of cells (m x n).

4. *Interpretation of the Results*

The comparison between actual road traffic and LP solutions, given in Table 1, shows fairly high correlations except for the categories "miscellaneous", and "coal and coke." In general, the closer the commodity class is to representing a single good, the better the correspondence. The exception is the case of "coal and coke." Most of the movement of this class of goods in the economy is in fact by rail, so the pattern of road movements is likely to present an erratic picture.

The m+n-1 flows of the LP solutions are at considerable variance with the wider dispersal of actual flows. This is also indicated by the parameters of the regression equations. It arises because of the crudeness of the commodity groups and the coarseness of the zoning. We have seen that the model replicates a perfectly competitive equilibrium for trade in a homogeneous good, thus prohibiting cross hauls, i.e., movement of the good in both directions between any pair of regions. If the commodity class masks some significant product differentiation, e.g., that between coking coal and lignite, then the assumption of the model is violated and, in reality, cross hauls and a more diffuse distribution might not be irrational. Similarly, if the zones are very large and the good to be

Commodity	r^2	Intercept	Slope	% Cells Occupied in Actual Matrix
1	0.39	541.3	0.54	18
4	0.87	52.4	0.70	9
5	0.89	−87.1	1.41	37
6	0.69	−45.8	1.03	20
7	0.79	14.1	0.56	40
8	0.82	402.2	0.62	40
9	0.51	−32.8	1.29	33
10	0.81	149.3	1.04	41
11	0.57	−94.3	1.15	61
12	0.82	−61.4	1.18	35
13	0.20	295.0	0.28	41

m+n−1 as a percentage of total cells 2.5%

TABLE 1.   COMPARISON OF ACTUAL ROAD FLOWS WITH LP SOLUTIONS

distributed is produced and consumed in several locations within regions, then the assumption of a single transport cost does not hold and quite rational cross hauling between zones might occur. The disparity is greater in the case of the heterogeneous categories "other manufactures," "miscellaneous," "transport equipment," and "building materials," and is much less with the more uniform mineral traffics.

Table 2 shows the comparisons for the total interregional movement of commodities 1 to 6. For categories 1, 4, 5, and 6 the correlations are far higher, even if the actual dispersal is a little wider, when rail traffic is included in the calculations. Categories 2 and 3, "iron-ore" and "limestone," travel exclusively on the rail system and the results for these are most gratifying--with high correlations, i.e., intercept values close to zero and slopes close to unity. The percentages of occupied cells in the actual origin/destination matrices are very low, as many of the 78 regions do not generate or attract these traffics. For "iron-ore" in the LP solution, m+n−1 = 37 (0.5% of cells) and for "limestone" m+n−1 = 43 (0.2% of cells).

Table 3 reveals that actual and LP flows of commodities by road between the 24 major cities compare much more favorably than those between all 78 regions with respect to correlations and the regression equations. Performance still varies with homogeneity of the commodity classes. In each table r^2 is the correlation coefficient, i.e., the closer it is to 1 the higher the correlation.

Commodity	r^2	Intercept	Slope	% Cells Occupied in Actual Matrix
1	0.85	-857.9	1.14	26
2	0.91	3.5	0.99	1
3	0.89	-14.1	1.04	2
4	0.82	-38.4	1.24	18
5	0.88	-96.9	1.42	45
6	0.70	-86.6	1.16	21

For 1, 4, 5, and 6 m + n - 1 as a percent of total cells 2.5%

For 2   m + n - 1 as a percent of total cells 0.5%

For 3   m + n - 1 as a percent of total cells 0.2%

TABLE 2.   COMPARISON OF TOTAL ACTUAL FLOWS WITH LP SOLUTIONS

Commodity	r^2	Intercept	Slope	% Cells Occupied in Actual Matrix
1	0.99	-128.5	1.03	27
4	0.98	-43.3	1.08	21
5	0.96	-140.7	1.21	53
6	0.98	-152.7	1.11	26
7	0.94	-20.8	1.23	58
8	0.96	-524.5	1.18	58
9	0.91	-139.3	1.27	48
10	0.96	-247.4	1.07	49
11	0.49	-463.3	1.33	77
12	0.97	-138.6	1.16	45
13	0.97	-253.6	1.17	56

m + n - 1 as a percentage of total cells 8.3%

TABLE 3.   COMPARISON OF ACTUAL ROAD FLOWS BETWEEN 24 MAJOR CITIES WITH
            LP SOLUTIONS

Dual Variables

     In embarking on a description and interpretation of the dual variables, some
major difficulties due to the nature of the data must be borne in mind.

     Because the Ministry of Transport survey was essentially of vehicle move-
ments, the trips which were summed to give tonnages in these origin/destination
matrices might only represent part of total journeys from production locations to

places of consumption. Many of the trips represent stages between transship-
ments--movements to, and distribution from ports, depots, warehouses, rail heads,
etc. The movements reflect the transport sector's operations more closely than
the geography of production and consumption. They are transport movements, not
the total paths of output and input consignments. From the point of view of the
examination of dual values, the combination of wide variations in the area, mix of
activities, and population density of the zones is an additional weakness, because
the ratio of intra- to inter-zonal movement can vary with any or all of these in
a very uncertain fashion.

Examining the dual values for road movements alone is a dangerous enterprise,
since, for many commodities, there is a significant movement by rail and a not
altogether clear relationship between length of trip and mode used. It would be
tedious, cumbersome and of dubious value to present the dual values for road
movements here and attempt interpretations of their geographical variations.

When the dual values for total road plus rail movements are considered, at
least some of the distorting effects are overcome. Also, the greater correspondence of actual and LP flows in these cases is cause for greater confidence in
making interpretations.

The range between the highest and lowest shadow price for each commodity was
divided into quartiles. The shadow price values were then mapped by shading each
origin or destination zone according into which quartile the shadow price
associated with it fell. Some of the distributions of values in Figures 1 to 12
do make geographical sense and there is a healthy clustering of values for zones
of similar locational and/or activity characteristics, despite disparities in
size. For coal and coke, the transport system operations reflected in the flows
confuse the picture, since the traffic is originating from distribution points as
well as coalfields, but for both origins and destinations prices are high in areas
remote from major coalfields, i.e., southern England, and decrease toward the
sources of coal. The values for iron ore at origins and destinations do increase
generally with distance from the domestic sources around Northampton, but the
movement of imported ores complicates matters. Sources of limestone are quite
widespread in Great Britain and the shadow prices vary with general accessibility.
In central areas, origin and destination prices are low, whereas in peripheral
areas they are high. Scrap prices seem to be higher at origins with heavy
demand--steel producing areas. Supply areas in southern England with little steel
production have low values. Prices for steel are low at production points and
high in consuming areas. The picture presented by the values of oil is the more
heartening justification of this effort. For this fairly homogeneous commodity
moving from a few coastal importing and refining points to all regions of the
country, low prices at these origins and the increase in origin price with

302   P. O'SULLIVAN

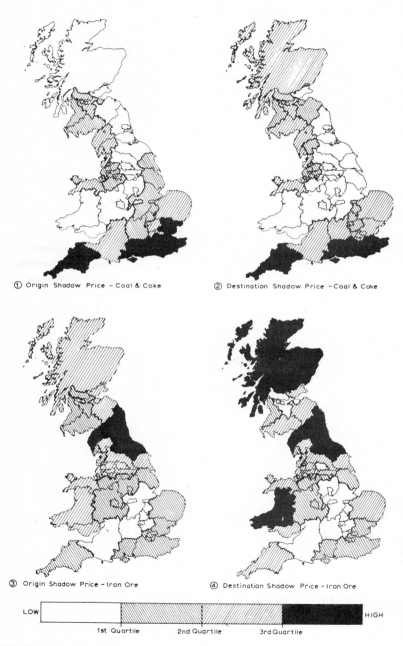

① Origin Shadow Price – Coal & Coke   ② Destination Shadow Price – Coal & Coke

③ Origin Shadow Price – Iron Ore   ④ Destination Shadow Price – Iron Ore

LOW |_____|////////|////////|████████| HIGH
         1st Quartile   2nd Quartile   3rd Quartile

FIGURES 1 – 4.

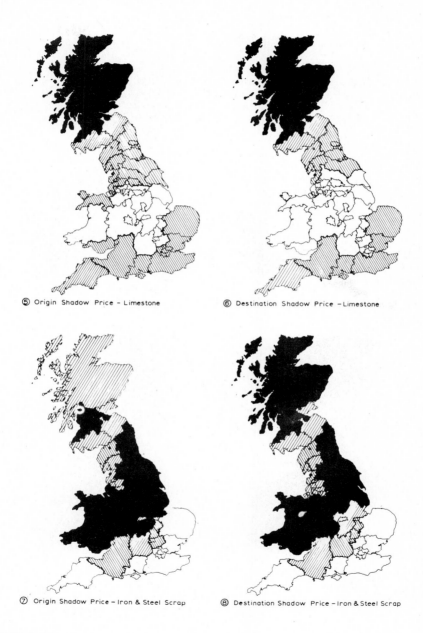

⑤ Origin Shadow Price – Limestone

⑥ Destination Shadow Price – Limestone

⑦ Origin Shadow Price – Iron & Steel Scrap

⑧ Destination Shadow Price – Iron & Steel Scrap

FIGURES 5 – 8.

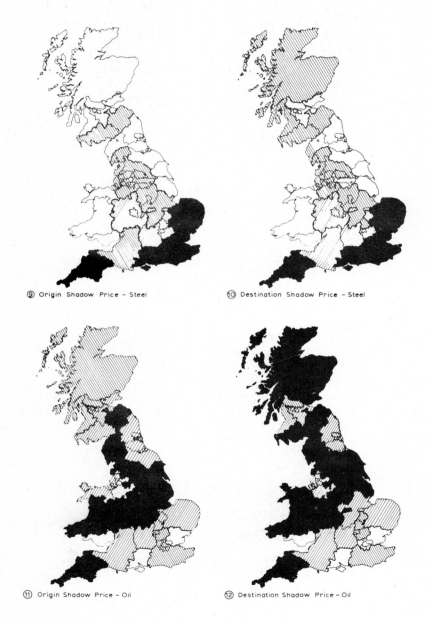

⑨ Origin Shadow Price – Steel

⑩ Destination Shadow Price – Steel

⑪ Origin Shadow Price – Oil

⑫ Destination Shadow Price – Oil

FIGURES 9 – 12.

distance from them seems sensible. The same pattern holds for prices at desti-
nations--the more remote the area from the major refinery locations, the higher
the price.

5. *Conclusion*

It is evident from the results that the predictions of the model are better
for commodity categories that are closer to representing a homogeneous good, when
road and rail traffic are distributed jointly, and when the zones between which
traffic is allocated are more homogeneous with respect to size and density of
activity.

To develop the first point, it was hoped that the results of a 1967/1968
road freight survey would provide data on tonnages originating and terminating in
zones for 30 commodity classes and that allocating these using LP would make more
sense and provide a better fit to reality. However, the survey results were not
readily forthcoming and we abandoned the scheme.

On the matter of the joint allocation of road and rail traffic, the point is
that we saw our project largely as an exercise in forecasting interregional trade
and it appears that a better estimate could be produced when the modal split of
the traffic is ignored. There was really no adequate description of the relative
cost and performance characteristics of the two modes at the scale at which we
were working (we had to use road distances as a surrogate for transport costs)
and, thus, any attempt to allocate traffic simultaneously between modes and
regions would be futile. The general implication of the results given here for
modelling the freight transport economy, in terms of predictability, is that we
should estimate the total volume of movement first before proceeding to allocate
traffic to different modes of transport. We did examine the modal split revealed
by the data on actual flows (Ref. 2, Chapter 7) but no simple relationship with
respect to our transport cost proxy, distance, emerged. With the data we had we
were unable to generalize about modal split either in terms of statistical
regularity or optimal allocation. In fact, rail is only a significant carrier for
a limited number of classes of goods, namely, limestone, iron-ore, coal and coke,
and scrap and steel. In all other categories rail accounts for less than one-fifth
of the tonnage.

The implication of the third point is that the use of a more homogeneous set
of zones, in terms of size at least, might improve predictability. The better
predictions of inter-urban flows prompts an examination of the pattern of actual
flows for the existence of hierarchical structure, i.e., relationships between
zones being based on the rank order of economic importance of their central
cities. If this revealed a clear hierarchical network of flows for some

commodities, it would seem sensible to distribute traffic on this basis, solving
for flows between major cities while dealing with their transactions with their
surrounding regions separately.

     A major drawback to using LP for predicting at this scale is the fact that
the m + n - 1 solution values for the most part failed to give anything like the
dispersal of actual traffic. It is noteworthy that for the most dispersed, heterogeneous commodity class (miscellaneous) a gravity model provides a better fit to
reality. The results of an earlier gravity model fitting exercise (Ref. 7) are
displayed in Table 4. Comparison of Tables 1 and 4 suggests an inverse complemen-
tarity between the two methods; that is, the better the one, the worse the other.

Commodity	1	4	5	6	7	8	9	10	11	12	13
r^2	0.49	0.28	0.24	0.37	0.41	0.62	0.31	0.60	0.43	0.45	0.52

TABLE 4.   COMPARISON OF ACTUAL ROAD FLOWS WITH GRAVITY MODEL ESTIMATES

     In all, it was gratifying that better predictions were achieved with the LP
model rather than the gravity formulation. It is more satisfying to sustain the
assumption of even a local tendency to optimality in the economy, rather than the
gravity model assumption that statistical regularities observed at one point in
the past will persist inevitably. The results presented here suggest that there
is little doubt but that LP solutions with a more homogeneous set of zones and a
finer commodity categorization will predict freight flows better than the gravity
model. A further advantage lies in the lead into investigation of regional comparative cost advantage variations provided by the dual shadow prices. However,
presently we are in a limbo where models must be chosen according to the speci-
ficity and accuracy of the information available. In many instances, where the
data represents an aggregation accross a wide spectrum of specific needs, the
gravity model is the most reasonable device developed to produce the quantitative
estimates which have to be hazarded.

     The specific results of this study have not been used for planning purposes
in Great Britain, however, the general finding in favor of LP as a forecasting
model for homogeneous goods traded between small regions did encourage its application in a regional road investment planning exercise (Ref. 1). In order to
program investment in the road network of a province of Indonesia, estimates of
future trade in agricultural staples between sub-regions of the province were
produced by solving transportation problems to reconcile deficits and surpluses
calculated from projections of yield and consumption. The resulting estimates of

demand for transport were then input to a mixed integer programming formulation
which determined the network additions and improvements that could be financed by
a given budget and minimized total vehicle operating costs throughout the system.

References

1. Ahmed, Y., P. O'Sullivan, Sujono, and D. Wilson; "Road Investment Programming
for Developing Countries: A Case Study of S. Sulawesi", Transportation Center
Report, Northwestern University (forthcoming).

2. Chisholm, M., and P. O'Sullivan; Freight Flows and Spatial Aspects of the
British Economy, Cambridge University Press, London, 1973.

3. Dorfman, R., P. Samuelson, and P. Solow; Linear Programming and Economic
Analysis, McGraw Hill, New York, 1958.

4. Dunford, M.; "Interregional Freight Flows: An Application of Linear
Programming", B.Sc. Dissertation, Department of Geography, University of
Bristol, England, 1969.

5. Ministry of Transport; Survey of Road Goods Transport 1962, Statistical Paper
6, H. M. S. O., London, 1966.

6. Ministry of Transport; Survey of Road Goods Transport 1962, The Magnetic Tapes
from the Geographical Analysis and Their Use in the Transport Cost Model,
Mathematical Advisory Unit Note 101, 1968.

7. O'Sullivan, P.; "Forecasting Interregional Freight Flows," Colston Symposium
on Regional Forecasting, Bristol, 1970.

8. Stevens, B.; "Linear Programming and Location Rent," Journal of Regional
Science 3, 15-26 (1961).

9. Wilson, A.; Entropy in Urban and Regional Modelling, Pion, London, 1970.

Chapter 15

NEW COMMUNITY DEVELOPMENT WITH THE AID
OF LINEAR PROGRAMMING

Richard L. Heroux
Centennial Corporation
Centennial Park
Grand Rapids, Michigan 49501

and

William A. Wallace
Rensselaer Polytechnic Institute
Troy, New York   12181

Abstract. A linear programming model is presented
as a tool for dealing with the complexities of
large scale land developments or new communities.
It has been used for planning a new community in
upstate New York. An illustrative example of a
development for 100,000 people is included.

1. *Introduction*

The creation of new communities is a means of alleviating the ills of urban-
ization, as noted by Title VII of the Federal Housing Act of 1970 and the formation
of state community development authorities such as the New York State Urban
Development Corporation. Programs such as these seek to provide living environments that do not include spiralling land cost, costly public facilities, increased
cost of and inaccessibility to transportation, poor accessibility to schools and
shopping centers, inefficient county government and wasted land resources, but do
include suburban housing for low income and minority groups. The purpose of this
chapter is to illustrate how linear programming can be used to determine the most
economical way of developing a new community while meeting living requirements and
social objectives.

2. *The Model*

Notation

Knowns or input parameters

N_{ij} is the number of apartments in a building or house of type i constructed
on a lot of size j;

Q_i is the average number of people who will live in each type i dwelling;

P_i is the maximum population projected for each type i dwelling, independently
of the lot size;

A   is total acreage available for development;

E   is an estimate of the total number of the projected population that will
require employment;

M   is an estimate of the total number of people with employment needs which
can be satisfied outside the planned development;

δ_t is the fraction of the labor force which is employable in a type t industry;

r_k is the estimated number of school students of type k (elementary, junior high,
etc.) per dwelling, independent of the type of housing;

G_t is the average number of people per dwelling employable in a type t
        industry;

D_k is the maximum student population projected for each school type k;

$a(\cdot)$ is the per unit acreage requirements for each type of land use defined by
        $(\cdot)$;

a_{ij} is the number of acres required for a type i dwelling on a lot of size j;

K   is the aggregate development cost per dwelling;

p   is the average family size per dwelling.

Unknowns or decision variables

X_{ij} is the number of houses or dwellings of type i which will be constructed
        on lots of size j;

$Z(\cdot)$ is the proportion of total available land used for uses defined by $(\cdot)$;

Z_i is the proportion of available land used for housing of type i;

Y   is the total number of dwellings constructed; i.e., $Y = \sum_i \sum_j N_{ij} X_{ij}$.

Constraints

The type of community is dependent upon the nature of the constraints imposed,
voluntarily or otherwise, during the development process. These constraints
brought about by market forces, zoning, etc., coupled with the developer's plan for
the site, will result in the various land uses such as residential, commercial,
educational, recreational, etc., which give the new community its character. The
interrelationship among these various land uses may not be known initially, but
can be conceptualized for the development at the horizon. Using this conceptual
approach, a model is structured so that the complexities arising from the inter-
relatedness of the various elements of the development are accounted for.

Population Constraints. Factors which dictate the number of dwellings to be
built are the projected demand for various dwelling types based on market analysis,
the average number of people which typically occupy a particular dwelling type,
and the amount of land used in the construction of each type of dwelling. To
insure that a sufficient number of dwellings of type i are built, we must have

$$\sum_j Q_i X_{ij} \leq P_i \qquad \text{for each i.} \qquad (1)$$

In order that the community may be viable, due consideration must be given to
the availability of employment opportunities for the adults and educational
facilities for the children.

312   R. HEROUX AND W. WALLACE

Of the planned inhabitation, only a certain number E will require employment.
Employment opportunities within the surrounding area will satisfy M people's need.
Based on labor statistics, it is possible to determine the proportion of the remainder which will seek employment in a type t industry. The number of dwellings
constructed (Y) should allow all those seeking employment in industry of type t
$(G_t Y)$ to secure a position, and thus

$$G_t Y \leq \delta_t (E-M) \qquad \text{for each t.} \qquad (2)$$

In an analogous manner, once estimates are made of the number of children in
each family unit which are likely to require various levels of education (r_k), the
number of students requiring the kth level of education should not exceed the projected number of students planned for the kth type of educational facility (D_k), or

$$r_k Y \leq D_k \qquad \text{for each k.} \qquad (3)$$

Land Usage Constraints. The various land uses for a new community are highly
interrelated since the extent to which land is made available for each use is
dependent upon the total population planned for and the per capita acreage requirements of each of the uses.

The land available for development is known; however, the proportion of this
land which should be made available for each of the land uses in order to provide
for a well balanced community is not known. By determining this proportion
endogenously, we can be sure that a balanced community will result.

The total land required for each type of residential dwelling will depend on
the lot size used. Once a maximum density (number of houses per acre) has been
determined for each type (i) of housing, the land required for i^{th} type of dwelling on a lot of size j, (a_{ij}) is computed. Then to guarantee that residential
land requirements do not exceed total land availability, we have

$$\sum_j a_{ij} X_{ij} \leq Z_i A \qquad \text{for each i,} \qquad (4)$$

where Z_i is the proportion of the available land, A, used for housing of type i.
In general, the relationship between non-residential land requirements and
total land availability is

$$a(\cdot)Y \leq Z(\cdot)A \qquad \text{for each land use } (\cdot), \qquad (5)$$

where $a(\cdot)$ can always be reduced to acres per dwelling irrespective of the land
use. Each of the land uses represented by the dot notation $((\cdot))$ is described

below and produces constraints of type (5). In each case, a($\cdot$) is in acres per
dwelling.

(a) *Industry*

Industrial land requirements are determined by computing the average number of
employees per acre for each type of industry. The product of a_t (acres per
employee for industry of type t) and the number of people per dwelling employable
in industry of type t, G_t, gives the industrial land requirements (in acres per
dwelling) for each type of industry.

Therefore, for industrial land usage, a($\cdot$) is replaced by a(t), where for each
industry of type t, a(t) = $a_t G_t$. Similarly, Z($\cdot$) in (5) is replaced by Z_t, where
Z_t represents that proportion of the total land used for industry of type t. The
constraints of type (5) are

$$a(t)Y \leq Z(t)A$$

or $$a_t G_t Y \leq Z_t A \qquad \text{for each t.}$$

Similarly, analogous expressions for a($\cdot$) and Z($\cdot$) are calculable for the following
land uses.

(b) *Educational Institutions*

Educational institutions are required in the planned development in order to
serve the educational needs of the student population (and attract families as
residents). The land required by each institution, k, is approximated by using
existing standards. They include acreage requirements a_k for an institution of
type k, the number of students n_k per institution of type k, and the number of
school students of type k per dwelling, r_k. In this case, a(k) = $a_k r_k / n_k$ for
each k.

(c) *Commercial*

Determining the amount of commercial building space requires an estimate L_c
of the share of the market which can be expected to be captured by each type c of
commercial establishment being considered (measured as the number of people per
establishment). In addition, a knowledge of the average family size per dwelling
p, and the acreage requirements for each type of establishment, a_c, is needed.
Here, a(c) = $p a_c / L_c$ for each c.

(d) *Public Services*

Land requirements for public services are defined in terms of public service
building space. Parameters to be specified are the average family size p per
dwelling, acreage requirements a_q per public service building of type q, and the
number of public service buildings n_q of type q per 1000 population. In our

notation, $a(q) = 1000pa_q n_q$ for each q.

(e) *Health Services*

Land requirements for health services are defined in terms of health facility needs. They include the average family size p per dwelling, acreage requirements a_h for a health facility of type h, and the number of health facilities n_h per 10,000 population. Or, $a(h) = 10,000pa_h n_h$ for each h.

(f) *Recreational Facilities*

Recreational land requirements are a function of the average family size p per dwelling, the acreage requirement a_f for a facility of type f, and the number of recreational facilities n_f per 1000 population. Therefore, $a(f) = 1000pa_f n_f$ for each f.

(g) *Cultural Facilities*

Land for cultural facilities is dependent on the standard acreage requirement a_ℓ for a facility of type ℓ, the number n_ℓ of type ℓ facilities per 1000 population, and the average family size p per apartment. Or, $a(\ell) = 1000pa_\ell n_\ell$ for each ℓ.

(h) *Cemeteries*

In a large scale planned development, land must be set aside for cemetery usage. Standards are available giving the acreage requirements a_g per 1000 population. Also required is the average family size p per dwelling; thus, $a(g) = 1000pa_g$ for each g.

(j) *Churches*

The land needed for places of worship depends on the average family size p per dwelling, the acreage required a_m for each church of type m, the standard number n_m of churches of type m per 1500 people, and the proportion α of the population to be served by places of worship. Therefore, $a(m) = 1500\alpha pa_m n_m$ for each m.

Finally, the balance of the community is assured by constraining the proportions of land devoted to each type of land use to sum to one; or

$$\sum_i Z_i + \sum_t Z(t) + \sum_k Z(k) + \sum_c Z(c) + \sum_q Z(q) + \sum_h Z(h) + \sum_f Z(f) +$$

$$\sum_\ell Z(\ell) + \sum_g Z(g) + \sum_m Z(m) = 1, \tag{6}$$

where in $Z(\cdot)$, $(\cdot)$ assumes the subscripts in (6) and each identifies the respective land uses enumerated.

Other Constraints. It may be noted that several infrastructure elements of the planned development are missing, notably transportation, utilities, and open space. In this formulation, it is assumed that the land allocated to roads is a fixed multiple of the land used for all other elements determined by constraints (4) and (5). The transportation constraints are therefore redundant. Also, the land used for a sewage treatment plant and a reservoir are known and constant.

Utilities, on the other hand, consume no additional land, for it is assumed they follow the roads. Consequently, utility constraints which are expressed in feet would be identical to the transportation constraints except for a factor which converts miles to feet. Therefore, these constraints are redundant as well.

The Objective Function

It is assumed that the developer's objective is to maximize the profit before infrastructure and financing costs. This objective is reasonable since some of these costs are often assumed by the municipality. When this is not the case, these costs must be met by the developer and must be deducted from the total maximum net revenue to decide on the real economics of the project. In this chapter, however, we have considered the costs associated with financing, roads, utilities, a reservoir, a sewage treatment plant, and open space exogenous to the model.

The profit in land development is due to the sale of houses in the case of residential land development, and to construction and the sale of land for industrial and commercial uses. Other land uses are developed to support the sale of houses, and the construction and the sale of industrial and commercial land at a cost to the developer. In order to determine the profit, it is necessary to identify or estimate the unit costs and revenues associated with each element in the development plan. The developer's objective is therefore, to maximize

$$\sum_i \sum_j P_{ij} X_{ij} + KY, \tag{7}$$

where P_{ij} denotes the profit on a dwelling of type i built on a lot of size j, Y is the number of units constructed, and K is the profit per dwelling. In our notation,

$$K = \sum_t p_t a(t) + \sum_k c_k r_k + \sum_c p_c a(c) + \sum_q c_q a(q) + \sum_h c_h a(h)$$

$$+ \sum_f c_f a(f) + \sum_\ell c_\ell a(\ell) + \sum_g c_g a(g) + \sum_m c_m a(m),$$

and,

p_t is the profit per acre for industrial land;

p_c is the profit per acre for commercial land;

$-c_k$ is the profit per student;

$-c_q$ is the profit per acre for service buildings;

$-c_h$ is the profit per acre for health facilities;

$-c_f$ is the profit per acre for recreational facilities;

$-c_\ell$ is the profit per acre for cultural facilities;

$-c_g$ is the profit per acre for cemetery use;

$-c_m$ is the profit per church.

It should be noted that some of the variables in the linear programming model (1)-(6) are integer; namely, the X_{ij}'s. This suggests that a mixed integer code be considered. However, the magnitude of these variables is sufficiently large to justify the use of linear programming.

3. An Example

To illustrate the model, an example of a development for 100,000 people is presented. It is assumed that this development will be within the economic sphere of a metropolitan area; therefore, it need not be completely self contained. Further, for ease of exposition, it is assumed that the developer is also the builder and retails all property, elements characteristically rented (shopping centers, apartments, etc.) are sold, and a commitment by public officials has been made to provide services such as health clinics, a hospital, fire and police protection, a reservoir, and a sewage treatment plant. Also, the developer believes that to improve the marketability of saleable property within the development it is necessary to provide churches, recreational areas, public parks, schools, cultural facilities, cemeteries, and open space.

Data Requirements

The information for this example is based on previous studies (Ref. 4, 5, and 6). In addition, interviews were conducted with representatives of the New York State Office of Planning Services and the Urban Development Corporation. Costs figures and data for the constraint coefficients are in Reference 3.

In this example, it is assumed that (i) of 11,147 acres of total available land, 20% or 2,229 acres must be set aside for road construction; (ii) 3,139 acres is set aside for open space, 4.6 acres for a reservoir, and 10 acres for a sewage treatment plant, leaving 5,763.4 acres to be endogenously allocated by the model; (iii) the municipality will incur the full expense associated with roads and utilities; and (iv) the cost of financing will be 6% of all investments (see Table 5).

Results

The dimensions of this problem are 28 constraints by 37 decision variables and were solved in less than 1 minute using the MPS/360 package on an IBM 360/50 computer. The data is exhibited in Table 1.

The land use plan generated for the development (see Table 2-columns (3) and (4), and Table 3) gives the optimal allotment of land for the various uses. (Figures are rounded off to the nearest integer.)

With an understanding of the implications and interactions characteristic to any particular land use plan, the plan may be modified as given or may serve as a check for changes made by the parties involved. Proposed changes in land use are easily conducted and evaluated within the framework of the model. For instance, $Z(v)$ gives the proportion of the total available land (A) which is used for a particular use, v. If, for a specified use, the land allocated to it (i.e., $Z(v)(A)$ is judged to be too great, it can be reduced using parametric linear programming.

Noticeably missing from the optimal land use plan were dwelling types X_{12}, X_{21}, X_{22}, X_{31}, and X_{32}. Those dwelling types which are in the optimal solution suggest the most favorable dwelling mix and density to employ. In contrast, those types which the model indicates should not be built reveal the least profitable dwelling types and the degree to which each would decrease the profit if they were forced into the solution. In this example, if the developer, say for design purposes, added any of the dwelling types X_{12}, X_{21}, X_{22}, X_{31}, and X_{32}, each would reduce the profit by $89, $6323, $2905, $1960, and $1230, respectively. Also, since these dwellings are considered unfavorable in a profit maximization sense, the corresponding average lot sizes associated with each dwelling type is unattractive as well.

The population distribution for the optimal land use plan is shown in Table 4. The dwelling mix, densities, and population distribution given are not explicit guidelines but serve as an aid to the developer.

4. *Conclusion*

Although not constructed specifically to assist land developers, the model was used for planning a prospective new community (The Highlands Project) in upstate New York (Ref. 7). The linear programming formulation was extended to include multi-periods in order to analyze the financial feasibility of the

No.	Type (Constraints)	Indices		Coefficients			Unit Profits	Bounds — Unknown	Bounds — Known
1.	Housing (1)	i=1	j=1	a_{ij}=-.137	Q_i=-3.6	N_{ij}=-1	P_{ij}=-3000	Z_1^A	P_i=-69,560
2.			=2	=-.171	=-3.6	=-1	=-3310	Z_1^A	=-69,560
3.			=3	=-.206	=-3.6	=-1	=-3810	Z_1^A	=-69,560
4.		=2	=1	=-.400	=-50.4	=-14	=-2200	Z_2^A	=-15,220
5.			=2	=-.562	=-50.4	=-14	=-2600	Z_2^A	=-15,220
6.			=3	=-.720	=-50.4	=-14	=-2940	Z_2^A	=-15,220
7.		=3	=1	=-.046	=-18.0	=-5	=-1800	Z_3^A	=-15,220
8.			=2	=-.069	=-18.0	=-5	=-2000	Z_3^A	=-15,220
9.			=3	=-.092	=-18.0	=-5	=-2300	Z_3^A	=-15,220
10.	Industry (2)	t=4		G_t=-.216	a_t=-1/16		P_t=-365,000	Z_4^A	E=-40,000 M=-20,000 δ_t=-30
11.	Schools (3)	k=5		r_k=-60	a_k=9.5	n_k=600	$-c_k$=-2034	Z_5^A	D_k=-16,800
12.		=6		=-.46	=30	=-1740	=-3334	Z_6^A	=-12,880
13.		=7		=-.07	=100	=-1800	=8900	Z_7^A	=-1,800
14.	Commercial (5)	c=8		a_c=4/1000	L_c=-.8	p=-3.6	p_c=-301,000	Z_8^A	
15.		=9		=-3.75/1000	=-.2	=-3.6	=-472,000	Z_9^A	
16.	Public services (5)	q=10		n_q=-.13/1000	a_q=-1.5	p=-3.6	$-c_p$=-104,000	Z_{10}^A	
17.		=11		=-.04/1000	=-5.0	=-3.6	= 6,500	Z_{11}^A	
18.	Health services (5)	h=12		a_h=-1	n_h=-1/10,000	p=-3.6	$-c_h$=-1,984,000	Z_{12}^A	
19.		=13		=40	=-1/10,000	=-3.6	= 400,000	Z_{13}^A	
20.	Recreational areas (5)	f=14		a_f=-.5/1000		p=-3.6	c_f=-10,000	Z_{14}^A	
21.		=15		=-1.5/1000		=-3.6	=-10,000	Z_{15}^A	
22.		=16		=-2.0/1000		=-3.6	=-10,000	Z_{16}^A	
23.		=17		=-1.5/1000		=-3.6	=-10,000	Z_{17}^A	
24.		=18		=-3.5/1000		=-3.6	=-10,000	Z_{18}^A	
25.	Cultural facilities (5)	ℓ=19		a_ℓ=-2.4	n_ℓ=-.01/1000	p=-3.6	$-c_\ell$=-1,750,000	Z_{19}^A	
26.		=20		=-1.38	=-.05/1000	=-3.6	=-4,020,000	Z_{20}^A	
27.	Cemeteries (5)	g=21		a_g=-.26/1000		p=-3.6	$-c_g$= 4,000	Z_{21}^A	
28.	Churches (5)	m=22		a_m=-2.06 =-.60	n_m=1/1500	p=-3.6	$-c_m$=-1,449,000	Z_{22}^A	

TABLE 1. BASIC LINEAR PROGRAMMING DATA

Index of Land Uses v =	Land Uses (1)	Proportion of total available land used for use v (α_v) (2)	Land used for items in column (1) in acres ($\alpha_v A$) (3)	Number of items in column (1) built (4)
	Dwellings			
1	X_{11}	⎰.55213	741	5409
	X_{13}	⎱	2441	11850
2	X_{23}	.03773	217	302
3	X_{33}	.01350	78	846
4	Mfg. and Whs.	.06023	347	na*
	Schools			
5	Elementary	.04239	244	23
6	Junior-senior high	.03539	204	7
7	Community college	.01735	100	1
	Commercial Establishments			
8	Neighborhood shopping center	.05140	296	4
9	Town shopping center	.01205	69	1
	Public Services			
10	Fire/police/office buildings	.00313	18	12
11	Garages/depots	.00321	18.5	4
	Health Services			
12	Clinics	.00161	9.3	9
13	Hospitals	.00642	37	1
	Parks and Recreational Areas			
14	Tot lots	.00803	46	na
15	Playgrounds	.02409	139	na
16	Neighborhood parks	.03212	185	na
17	Playfields	.02409	139	na
18	Community parks	.05622	324	na
	Cultural Facilities			
19	Performing arts center	.00039	2.2	1
20	Libraries	.00111	6.4	5
21	Cemeteries	.00418	24	na
22	Churches/houses of worship	.01324	76	37

*Does not apply.

TABLE 2.   OPTIMAL LAND USE PLAN FOR THE DEVELOPMENT

	Transportation (number of linear miles for (0))	Utilities		Land used for (3) (acres)
		(number of lineal feet for (1))	(number of (2) supplied)	
(0) Roads	769			
(1) Storm sewer system		4,059,046		
Water system		4,059,046		
Sanitary sewage system		4,059,046		
(2) Manholes			13,530	
Hydrants			8,118	
Valves			16,236	
(3) Sewage treatment plant				10
Reservoir				4.6
Open space				3139

TABLE 3.   INFRASTRUCTURE USAGE AS DETERMINED FROM THE OPTIMAL LAND USE PLAN

The total number of each type of people occupying a dwelling corresponding to X_{ij}	Dwelling Type			
	X_{11}	X_{13}	X_{23}	X_{33}
Residents	19,471	42,660	15,220	15,220
Employees	1,168	2,560	913	914
Elementary school students	3,245	7,110	2,537	2,537
High school students	2,488	5,451	1,945	1,945
College students	378	830	296	296
Density by dwelling type (number families per acre)	713	418	19.4	54.4

TABLE 4.   POPULATION DISTRIBUTION FOR THE OPTIMAL LAND USE PLAN

proposed development. The application of the extended model to the Highlands Project is discussed in Reference 3.

The maximum profit and return on total investment to the developer in this example is $144,591,311, and 11.1%, respectively. These figures are obtained from the Development and Municipal Account Summary (Table 5). It is of interest to note that if the municipality had not incurred the expense of either the roads or utilities, the developer would incur a substantial financial loss as a result of the undertaking. This point further substantiates the ample literature which refers to the plight of developers who would not attempt to undertake such an endeavor without substantial external aid.

New Profit from the LP Solution		$236,082,848

Investments Considered in
  Obtaining New Profit

Housing	$ 559,932,625
Industry	183,975,486
Commercial	211,558,063
Other*	249,460,045
Subtotal	$1,204,926,119

Other Investments

Reservoir	$ 1,000,000
Open space	6,279,400
Sewage treatment plant	10,830,000
Subtotal	$18,109,400

Cost of Financing (@ 6%)	$73,382,137

Total Investment	$1,296,417,756

True Profit

(Net profit – other expenses – financing cost)	$ 144,591,311

% Return on Total Investment	11.1%

MUNICIPAL ACCOUNT SUMMARY

Investments

Roads	$307,503,448
Utilities	450,012,846
Total	$757,516,294

* Refers to investments in schools, services, facilities, cemeteries, and
  churches.

TABLE 5.   DEVELOPMENT ACCOUNT SUMMARY

References

1. Alonso, W.; "The Mirage of New Towns," The Public Interest, 10-16, Spring
    1970.

2. Eichler, E. P., and M. Kaplan; The Community Builders, University of Califor-
    nia Press, Berkeley, California, 1967.

3. Heroux, R., and W. A. Wallace; Financial Analysis and the New Community
    Development Process, Praeger Publishers Inc., New York, 1973.

4. Stone, P. A.; Housing, Town Development, Land and Cost, Estates Gazette,
    London, 1964.

5. Wren, K.; "Financial Appraisal of a New City," Milton Keynes Development
    Corporation, Buckinghamshire, England, 1970.

6. The American City Corporation and the Rouse Company; "An Analysis of Develop-
    ment Trends and Projects and Recommendations for a New City in South
    Richmond," The Rouse Company, 1970.

7. "The Highlands Project: A Development/Conservation Plan for a New Community
    in the Capital District of New York State," Rensselaer Polytechnic Institute,
    Troy, New York, June 1971.

SOCIAL SCIENCE LIBRARY

Manor Road Building
Manor Road
Oxford OX1 3UQ
Tel: (2)71093 (enquiries and renewals)
http://www.ssl.ox.ac.uk

This is a NORMAL LOAN item.

We will email you a reminder before this item is due.

Please see http://www.ssl.ox.ac.uk/lending.html
for details on:

- loan policies; these are also displayed on the notice boards and in our library guide.

- how to check when your books are due back.

- how to renew your books, including information on the maximum number of renewals.
  Items may be renewed if not reserved by another reader. Items must be renewed before the library closes on the due date.

- level of fines; fines are charged on overdue books.

Please note that this item may be recalled during Term.